SEDIMENTS IN THE TEMA HARBOUR (GHANA): CHEMICAL POLLUTION AND SEDIMENTATION RATES

Benjamin O. Botwe

Thesis committee

Promotor
Prof. Dr Piet N.L. Lens
Professor of Environmental Biotechnology
IHE Delft Institute for Water Education, the Netherlands

Co-promotor
Prof. Elvis Nyarko
Professor of Marine Environmental Science
University of Ghana, Accra, Ghana

Other members
Prof. Dr Karen J. Keesman, Wageningen University & Research, the Netherlands
Prof. Kerstin Kuchta, Technical University of Hamburg, Germany
Dr Elin Vanlierde, Agentschap Informatie Vlaanderen, Brussels, Belgium
Dr Leonard Ost, Deltares, Delft, the Netherlands

This research was conducted under the auspices of the Graduate School for Socio-Economic and Natural Sciences of the Environment (SENSE)

Sediments in the Tema Harbour (Ghana): chemical pollution and sedimentation rates

Thesis
submitted in fulfilment of the requirements of
the Academic Board of Wageningen University and
the Academic Board of the IHE Delft Institute for Water Education
for the degree of doctor
to be defended in public
on Friday, 29 June 2018 at 01:30 p.m.
in Delft, the Netherlands

by Benjamin Osei Botwe
Born in Accra, Ghana

CRC Press/Balkema is an imprint of the Taylor & Francis Group, an informa business

Published by:
CRC Press/Balkema
Schipholweg 107C, 2316 XC, Leiden, the Netherlands
Pub.NL@taylorandfrancis.com
www.crcpress.com – www.taylorandfrancis.com

ISBN: 978-1-138-32351-3
ISBN: 978-94-6343-838-4
DOI: https://doi.org/10.18174/443802

To my children, Jayden, Gideon and Davida

Acknowledgements

I wish to express profound gratitude to the Dutch Government for funding this PhD study at the UNESCO-IHE Institute for Water Education in Delft through Nuffic in the framework of the Netherlands Fellowship Programme (NFP-PhD. 12/316). I also wish to thank the University of Ghana for the financial support provided through the Office of Research, Innovation and Development (ORID) under the Faculty Development Fund (UGFD/7/2012-2013/004). The logistical and technical assistance provided by the Ghana Ports and Harbours Authority is also acknowledged.

I am most grateful to Prof. dr. Piet N.L. Lens (Promotor) and Prof. Elvis Nyarko (Co-Promotor), whose excellent guidance and advice have brought me this far. The staff of the Department of Marine and Fisheries Sciences at the University of Ghana are acknowledged for their support.

I am also indebted to Alisa Mast and Gregory A. Wetherbee, both at the US Geological Survey, for donating a KB corer for my field sampling. I also wish to thank Jolanda Boots (Admissions Officer, UNESCO-IHE, Delft), who was instrumental in the technical matters of my PhD study at the UNESCO-IHE, Delft. Professor Ronny Blust is acknowledged for permitting me to conduct bioassay experiments at the Systemic Physiological and Ecotoxicological Research (SPHERE) laboratory of the University of Antwerp, Belgium. I received lots of encouragements from Frank and Ida van der Meulen and Cor Schipper for which I am grateful. My dear wife Janet Botwe, my brother Dr. Theophilus Botwe and my entire family are also acknowledged for their support. Furthermore, I am grateful to Mr. Bennet Atsu Foli for providing editorial assistance. Several others have contributed to the success of my PhD work in one way or the other, whose names are not mentioned.

Abstract

The Tema Harbour in Ghana has been in operation for nearly six decades and is subject to large influxes of sediments and sediment pollution due to the intense human activities in the harbour area. This thesis assessed sediment pollution in the Tema Harbour by using the standard 10-day *Corophium volutator* and 28-day *Hediste diversicolor* whole-sediment toxicity bioassays as well as chemical contaminant (DDTs, HCHs, PAHs and metal - Cd, Pb, Cr, Ni, Cu, Zn and As) data. The bioassay results showed significant *C. volutator* and *H. diversicolor* mortalities, indicating that the Tema Harbour sediments are polluted and toxic. Biota-sediment accumulation factors further revealed a high potential for bioaccumulation of the sediment-associated metals, which can have adverse implications for the food chain. Thus, the Tema Harbour sediments are unsuitable for disposal at sea without remediation.

The thesis further investigated sediment accumulation rates (SARs) in the Tema Harbour by the combined analyses of sediment trap and sediment core data. The sediment cores exhibited variable bulk density profiles, indicating highly dynamic and non-steady sedimentation conditions. ^{7}Be-derived gross-estimates of very recent sediment accumulation rates using the constant flux-constant sedimentation (CF-CS) model were in the range of 2.5-9.0 $g.cm^{-2}.y^{-1}$. These values were much lower than the estimated average settling fluxes from the sediment trap data (15.2-53.8 $g.cm^{-2}.y^{-1}$), indicating sediment resuspension plays an important role in the sedimentation process. Conventional ^{210}Pb sediment dating models did not allow any estimation of SARs in the Tema Harbour. The ^{210}Pb-based TERESA model, on the other hand, proved to be a good tool for quantifying sediment accumulation rates in the Tema Harbour with time-averaged values in the range of 1.4-3.0 $g.cm^{-2}.y^{-1}$ and sediment accretion rates of 1.7-3 $cm.y^{-1}$.

In conclusion, this study has shown that the Tema Harbour has been severely affected by anthropogenic activities, resulting in pollution of the sediments, especially those from the Fishing Harbour and the Canoe Basin. Moreover, the sediment accretion rates in the harbour may pose moderate problems for sustainable use of the harbour. There is, therefore, a need to improve sediment and environmental management in the Tema Harbour and regulate the disposal of the dredged material originating from this tropical coastal harbour.

Abbreviations and acronyms

AAS – Atomic absorption spectrometer

Ace – Acenaphthene

Acy – Acenaphthylene

AEDE – Annual effective dose equivalent

AGDE – Annual gonadal dose equivalent

AL1 – Action levels 1

AL2 – Action level 2

ANOVA – Analysis of variance

Ant – Anthracene

BaA – Benzo[a]anthracene

BaP – Benzo[a]pyrene

BbF – Benzo[b]fluoranthene

BCR – Community Bureau of Reference

BFR – Brominated flame retardant

BghiP – Benzo[g,h,i]perylene

BkF – Benzo[k]fluoranthene

Bq – Becquerel

BSAF – Biota-sediment accumulation factor

CANMET – Canada Centre for Mineral and Energy Technology

CB – Canoe Basin

Chr – Chrysene

COEC – Chemicals of emerging concern

CRM – Certified reference material

DahA – Dibenz[a,h]anthracene

DDD – Dichlorodiphenyldichlororethane

DDE – Dichlorodiphenyldichloroethylene

DDT – Dichlorodiphenyltrichloroethane

DNA – Deoxyribonucleic acid

DO – Dissolved oxygen

EF – Enrichment factor

E_h – Redox potential

ENEA – Ente per le Nuove tecnologie, l'Energia e l'Ambiente

ERL – Effects Range Low

ERLQ – Effects range low quotient

ERM – Effects Range Median

ERMQ – Effects range median quotient

ERICA – Environmental Risk from Ionising Contaminants Assessment and Management

Fla – Fluoranthene

Flu – Fluorene

GC – Gas chromatograph

GC-ECD – Gas chromatograph with electron capture detector

GC-FID – Gas chromatograph with flame ionisation detector

GC-MSD – Gas chromatograph with mass selective detector

GESAMP – Group of Experts on the Scientific Aspects of Marine Environmental Protection

GIS – Geographic Information Systems

GPHA – Ghana Ports and Harbours Authority

Gy – Gray

HCH – Hexachlorocyclohexane

H_{ex} – External hazard index

HPAHs – High molecular weight PAHs

IAEA – International Atomic Energy Agency

ICP – Inductively Coupled Plasma

ICP-MS – Inductively Coupled Plasma - Mass Spectrometry

IFH – Inner Fishing Harbour

I_{geo} – Geo-accumulation index

IP – Indeno[1,2,3-cd]pyrene

JICA – Japan International Co-operation Agency

LPAHs – Low molecular weight PAHs

MERMQ – Mean effects range median quotient

MH – Main Harbour

MTCA – Model Toxics Control Act

Nap – Naphthalene

NFP - Netherlands Fellowship Programme

NIST – National Institute of Standards and Technology

NTU – Nephelometric Turbidity Unit

NUFFIC - Netherlands Organisation for International Cooperation in Higher Education

OC – Organochlorine compound

OCP – Organochlorine pesticide

OFH – Outer Fishing Harbour

OM – Organic matter

PAH – Polycyclic aromatic hydrocarbon

PAH_{16} – 16 priority PAHs by the United States Environmental Protection Agency

$^{210}Pb_{ex}$ – Excess Lead-210

$^{210}Pb_{supp}$ – Supported Lead-210

PBDE – Polybrominated diphenyl ether

PCB – Polychlorinated biphenyl

PCDD – Polychlorinated dibenzodioxins

PCDF – Polychlorinated dibenzofuran

PCN – Polychlorinated naphthalene

PDBS – Phase Differencing Bathymetric Sonar

PFC – Perfluorinated chemical

PFOA – Perfluorooctanoic acid

PFOS – Perfluorooctane sulfonate

Phe – Phenanthrene

POPs – Persistent organic pollutants

ppm – Parts per million

PVC – Polyvinyl chloride

Pyr – Pyrene

RAC – Risk assessment code

Ra_{eq} – Radium equivalent activity

RHI – Radiological hazard indices

RSD – Relative standard deviations

SAR – Sediment accumulation rate

SF – Settling fluxes

SPHERE – Systemic Physiological and Ecotoxicological Research laboratory

SPM – Suspended particulate matter

SPSS – Statistical Package for Social Sciences

SQG – Sediment quality guidelines

SRM – Standard reference material

Sv – Sievert

SWI – Sediment-water interface

$T_{1/2}$ - Half-life

TBT – Tributyltin

TEF – Toxic equivalency factor

TENORM – Technologically enhanced naturally-occurring radioactive material

TEQ – Total toxicity equivalence

TERESA – Time Estimates from Random Entries of Sediments and Activities

TEU – Twenty-foot equivalent unit

$^{234}Th_{exc}$ – Excess Thorium-234

TN – Total nitrogen

TOC – Total organic carbon

TOR – Tema Oil Refinery

UN – United Nations

UNSCEAR – United Nations Scientific Committee on the Effects of Atomic Radiation

USEPA – United States Environmental Protection Agency

Contents

Chapter 1

General Introduction

1.1. Background of this study

Coastal harbours are among marine environments highly vulnerable to chemical pollution as they tend to receive and accumulate pollutants from maritime activities and other human activities within their catchments, including urbanisation, industrialisation and agriculture (Petrosillo et al., 2009; Smith et al., 2009; Lepland et al., 2010; Mestres et al., 2010; Schipper et al., 2010; Nyarko et al., 2014; Romero et al., 2014). Coastal harbours are also prone to siltation as a result of the influxes of sediment-laden seawater (Senten, 1989; Leys and Mulligan, 2011) with subsequent deposition and accumulation of the sediments within the harbour basin under favourable hydrodynamic conditions (Smith et al., 2009; Lepland et al., 2010; Luo et al., 2010; Mestres et al., 2010; Schipper et al., 2010). Chemical pollution and high sediment accumulation rates (SARs) in harbours are major environmental issues as they pose a threat to harbour sustainability and result in adverse human health, ecological and socio-economic impacts (Syvitski et al., 2005; Van Rijn 2005; Casado-Martinez et al., 2006; Birch and Hutson, 2009; Lepland et al., 2010; Mestres et al., 2010; Schipper et al., 2010; Green and Coco, 2014).

The management of sediment pollution and accumulation in harbours commonly involves dredging (Burton, 2002) with subsequent disposal in other aquatic or terrestrial ecosystems (Caille et al., 2003; Witt et al., 2004; Bolam et al., 2006; Schipper et al., 2010). Dredging has both economic and ecological implications as it is usually costly (Qu and Kelderman, 2001; Barneveld and Hugtenburg, 2008; Schipper et al., 2010) and moreover, the pollutants in the disposed materials are capable of exerting detrimental effects on the receiving ecosystems (Hong et al., 1995; Burton, 2002; Caille et al., 2003; Stronkhorst and van Hattum, 2003; Casado-Martinez et al., 2006; Birch and Hutson, 2009; Choueri et al 2009; Schipper et al., 2010). Thus, dredged material disposal is now recognized as a major environmental issue. Sediment pollutants of notable concern include metals and persistent organic pollutants (POPs), exemplified by polycyclic aromatic hydrocarbons (PAHs) and organochlorine pesticides (OCPs), due to their toxicity, environmental persistence and potential for bioaccumulation and transfer along food chains (Chau, 2005; Casado-Martinez et al., 2006; Schipper et al., 2010; Nyarko et al., 2011; Kelderman, 2012; Iqbal et al., 2013; Romano et al., 2013). Thus, information on sediment contamination and accumulation rates in harbours is relevant for their management, e.g., for predicting the time to dredge.

The coastal Tema Harbour in Ghana (Fig. 1.1) was constructed in 1960 and has been in operation since 1962 when it was commissioned. Industrialisation of the harbour area also started in the early 1960s and has seen much growth over the years.

Fig. 1.1 Aerial view of the Tema Harbour (Google Earth)

Major industrial activities in the harbour zone include oil refining, ore smelting, and manufacture of paints and cement. The harbour has fourteen berths including a dedicated oil berth, where oil is discharged by oil tankers, and a shipyard with two dry docks for servicing of ships. A wide range of ships call at the harbour including oil tankers, general cargo, container ships and bulk carriers, which transport e.g. fertilizer, ores, petroleum products and chemicals. Container traffic at the harbour increased by about 400%, i.e., from 150,000 Twenty-Foot Equivalent Units (TEUs) to 750,000 TEUs between the year 2000 and 2011 (http://ghanaports.gov.gh/GPHA). During this period, there were two major oil spills in the

3

Tema Harbour and a fire outbreak at the Tema shipyard, which destroyed oil pipeline installations at the harbour. Maintenance dredging is conducted in the harbour periodically with subsequent disposal in the adjacent sea. In order to increase the handling capacity of the Tema Harbour to about 1,000,000 TEUs, the Ghana Ports and Harbours Authority (GPHA) proposed to undertake expansion works at the Tema Harbour. It is expected that this project, when completed, will boost industrial activities and thus, the Tema Harbour may be highly impacted through important loads of pollutants and sediments. In order to understand and minimise the anthropogenic impacts, a comprehensive assessment of chemical pollution and sedimentation in the Tema Harbour is crucial.

Prior to this thesis, no comprehensive assessment of chemical pollution and sedimentation was conducted in the Tema Harbour. Baseline studies in the Tema Harbour have focused on total metal (http://open_jicareport.jica.go.jp/pdf/11681632_03.pdf; Nyarko et al., 2014) and total PAH (Gorleku et al., 2014) contamination in surface sediments. However, harbour sediments are often contaminated by a wide range of chemical pollutants (Casado-Martinez et al., 2006; Long et al., 2006; Birch and Hutson, 2009; Schipper et al., 2010). The aim of this thesis was, therefore, to investigate chemical pollution and sedimentation conditions in the Tema Harbour. The thesis objectives were to:

(1) Assess chemical contamination in sediments from the Tema Harbour and its ecotoxicological implication

(2) Evaluate the biological effects of chemical contamination in the Tema Harbour sediments

(3) Quantify settling fluxes of sediments and associated contaminants as well as recent sediment accumulation rates in the Tema Harbour.

1.2. Research approach

With an integral view of the research problem, a rigorous multidisciplinary and integrated approach was adopted to assess the chemical pollution and sedimentation in the Tema Harbour as shown in Fig. 1.2. To assess chemical pollution in the Tema Harbour sediments, chemical analysis and a screening-level risk assessment were first conducted on the Tema Harbour sediments to identify contaminants of potential concern by using Sediment Quality Guidelines (SQGs) and the Environmental Risk from Ionising Contaminants Assessment and Management (ERICA) tool. This screening-level risk assessment was reinforced by whole-sediment bioassays to evaluate the hazard potential of the Tema Harbour sediments and to assess the potential ecological impact of metal contamination in the harbour sediments.

To assess sedimentation conditions in the Tema Harbour, sediment core analysis was combined with the analysis of particulate matter accumulated in sediment traps at different periods. This allowed for an understanding of Suspended Particulate Matter (SPM) dynamics and the processes affecting the distribution and dynamics of sediment-associated contaminants in the Tema Harbour. The robust quantitative ^{210}Pb-based TERESA (Time Estimates from Random Entries of Sediments and Activities) model was combined with the measurement of short-lived radioisotopes (^{234}Th and ^{7}Be) in sediment cores to depict a reliable scenario for sedimentation in the Tema Harbour. This allowed for the quantification of recent SARs in the disturbed Tema Harbour where (1) conventional ^{210}Pb-based dating methods fail, (2) the use of sediment traps and ^{234}Th and ^{7}Be profiles in sediment cores show serious constraints, and (3) SARs fall beyond the capabilities of GIS-bathymetry.

1.3. Structure of this thesis

Before pursuing the thesis objectives, a review of pertinent literature was conducted and is presented in Chapter 2. The objectives of this thesis were pursued in Chapters 3-8. Chapters 3-5 dealt with objective 1: Chapter 3 focused on radionuclide contamination, Chapter 4 on organic pollutant (PAH, DDT and HCH) contamination, while Chapter 5 assessed the potential mobility of sediment-bound metals in surface sediments of the Tema Harbour and the ecological risk implications. The screening-level risk assessment was reinforced by whole-sediment bioassays to evaluate the hazard/toxic potential of the harbour sediments and assess the potential ecological impact of metal contamination in the sediments, and is presented in Chapter 6. Objective 3 is dealt with under chapters 7 and 8. Chapter 7 investigated SPM dynamics and recent sediment accumulation rates in the Tema Harbour, while Chapter 8 investigated settling fluxes of fine sedimentary metals and their ecotoxicological implications in the Tema Harbour. A synthesis of Chapters 3-8 is then presented in Chapter 9.

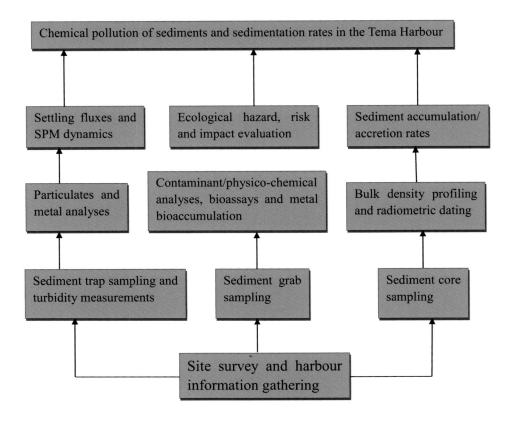

Fig. 1.2 Schematic presentation of the thesis approach to investigate chemical pollution of sediments and sedimentation rates in the Tema Harbour

References

Barneveld, H., & Hugtenburg, J. (2008). Feasibility study for implementation of sedimentation reduction measures in river harbours. RCEM, 1187-1192.

Birch, G.F., & Hutson, P. (2009). Use of sediment risk and ecological/conservation value for strategic management of estuarine environments: Sydney estuary, Australia. *Environ. Manage. 44*, 836-850.

Bolam, S.G., Rees, H.L., Somerfi eld, P., Smith, R., Clarke, K.R., Warwick, R.M., Atkins, M., & Garnacho, E. (2006). Ecological consequences of dredged material disposal in the marine environment: a holistic assessment of activities around the England and Wales coastline. Mar. Pollut. Bull. *52*(4), 415-26.

Burton, G.A. Jr. (2002). Sediment quality criteria in use around the world. *Limnology*, *3*(2), 65-76.

Caille, N., Tiffreau, C., Leyval, C. & Morel, J.L. (2003). Solubility of metals in an anoxic sediment during prolonged aeration. *Sci. Total Environ. 301*(1), 239-250.

Casado-Martínez, M.C., Buceta, J.L., Belzunce, M.J., & DelValls, T.A. (2006). Using sediment quality guidelines for dredged material management in commercial ports from Spain. *Environ. Int. 32*, 388-396.

Chau, K. W. (2005). Characterization of transboundary POP contamination in aquatic ecosystems of Pearl River delta. *Mar. Pollut. Bull. 51*(8–12), 960-965.

Choueri, R. B., Cesar, A., Abessa, D.M.S., Torres, R.J., Morais, R.D., Riba, I., Pereira, C.D.S., Nascimento, M.R.L., Mozeto, A.A., & DelValls, T.A. (2009). Development of site-specific sediment quality guidelines for North and South Atlantic littoral zones: Comparison against national and international sediment quality benchmarks. *J. Hazard. Mater. 170*, 320-331.

Gorleku, M.A., Carboo, D., Palm, L.M.N., Quasie, W.J. & Armah, A.K. (2014). Polycyclic aromatic hydrocarbons (PAHs) pollution in marine waters and sediments at the Tema Harbour, Ghana. *Acad. J. Environ. Sci. 2*(7), 108-115.

Green, M.O., & Coco, G. (2014). Review of wave-driven sediment resuspension and transport in estuaries. *Rev. Geophys*, 52(1):77-117.

Hong, H., Xu, L.-J., Zhang, L., Chen, J., Wong, Y. & Wan, T. (1995). Special guest paper: environmental fate and chemistry of organic pollutants in the sediment of Xiamen and Victoria Harbours. *Mar. Pollut. Bull. 31*(4), 229-236.

Iqbal, J., Tirmizi, S., & Shah, M. (2013). Statistical apportionment and risk assessment of selected metals in sediments from Rawal Lake (Pakistan). *Environ. Monit. Assess. 185*(1), 729-743.

Kelderman, P. (2012). Sediment Pollution, Transport, and Abatement Measures in the City Canals of Delft, the Netherlands. *Water Air Soil Pollut. 223*(7), 4627-4645.

Lepland, A., Andersen, T.J., Lepland, A., Arp, H.P.H., Alve, E., Breedveld, G.D., & Rindby, A. (2010). Sedimentation and chronology of heavy metal pollution in Oslo harbor, Norway. *Mar. Pollut. Bull. 60*(9), 1512-1522.

Leys, V., & Mulligan, R.P. (2011). Modelling coastal sediment transport for harbour planning: selected case studies. INTECH.

Long, E.R., Ingersoll, C.G. & MacDonald, D.D. (2006). Calculation and uses of mean sediment quality guideline quotients: a critical review. *Environ. Sci. Technol. 40(6)*, 1726-1736.

Luo, W., Lu, Y., Wang, T., Hu, W., Jiao, W., Naile, J.E., Khim, J.S., & Giesy, J.P. (2010). Ecological risk assessment of arsenic and metals in sediments of coastal areas of northern Bohai and Yellow Seas, China. *Ambio, 39*(5-6), 367-375.

Mestres, M., Sierra, J., Mösso, C., & Sánchez-Arcilla, A. (2010). Sources of contamination and modelled pollutant trajectories in a Mediterranean harbour (Tarragona, Spain). *Mar. Pollut. Bull. 60*(6), 898-907.

Nyarko, E., Botwe, B.O., Lamptey, E., Nuotuo, O., Foli, B.A., & Addo, M.A. (2011). Toxic metal concentrations in deep-sea sediments from the jubilee oil field and surrounding areas off the western coast of Ghana. *Tropical Environ. Res. 9*, 584-595.

Nyarko, E., Fletcher, A., Addo, S., Foli, B.A.K., & Mahu, E. (2014). Geochemical assessment of heavy metals in surface sediments: A case study of the Tema Port, Ghana. *J. Shipping Ocean Engin. 4*, 79-92.

Petrosillo, I., Valente, D., Zaccarelli, N., & Zurlini, G. (2009). Managing tourist harbors: Are managers aware of the real environmental risks? *Mar. Pollut. Bull. 58*, 1454-1461.

Qu, W., & Kelderman, P. (2001). Heavy metal contents in the Delft canal sediments and suspended solids of the River Rhine: multivariate analysis for source tracing. *Chemosphere, 45*(6), 919-925.

Romano, S., Piazza, R., Mugnai, C., Giuliani, S., Bellucci, L. G., Nguyen Huu, C., . . . Frignani, M. (2013). PBDEs and PCBs in sediments of the Thi Nai Lagoon (Central Vietnam) and soils from its mainland. *Chemosphere, 90*(9), 2396-2402.

Romero, A.F., Asmus, M.L., Milanelli, J.C.C., Buruaema, L., & Abessa, D.M.S. (2014). Self-diagnosis method as an assessment tool for environmental management of Brazilian ports. *Journal of Integr. Coastal Zone Manage. 14*(4), 637-644.

Schipper, C., Rietjens, I., Burgess, R., Murk, A. (2010). Application of bioassays in toxicological hazard, risk and impact assessments of dredged sediments. *Mar. Pollut. Bull. 60*(11), 2026-2042.

Senten, J.R. (1989) Pollution of harbour sediments by heavy metals. *Ocean & Shoreline Manage. 12*, 463-475.

Smith, J., Lee, K., Gobeil, C. & Macdonald, R. (2009). Natural rates of sediment containment of PAH, PCB and metal inventories in Sydney Harbour, Nova Scotia. *Sci. Total Environ. 407*(17), 4858-4869.

Stronkhorst, J., & Van Hattum, B. (2003). Contamination of concern in Dutch marine harbour sediments. *Arch. Environ. Contam. Toxicol. 45*, 306-316.

Syvitski, J.P, Vörösmarty, C.J, Kettner, A.J., Green, P. (2005). Impact of humans on the flux of terrestrial sediment to the global coastal ocean. *Science, 308*(5720), 376-380.

Van Rijn, L.C. (2005). Estuarine and coastal sedimentation problems. *Int. J. Sediment Res. 20*(1):39-51.

Witt, J., Schroeder, A., Knust, R., & Arntz, W.E. (2004). The impact of harbour sludge disposal on benthic macrofauna communities in the Weser estuary. *Helgol. Mar. Res. 58*, 117-128.

Chapter 2

Literature Review

2.1. Chemical pollution in coastal marine environments

The United Nations (UN) Group of Experts on the Scientific Aspects of Marine Environmental Protection (GESAMP) defined marine pollution as: "*The introduction by man, directly or indirectly, of substances or energy into the marine environment (including estuaries) resulting in such deleterious effects as harm to living resources, hazards to human health, hindrance to marine activities, including fishing, impairment of quality for use of sea water and reduction of amenities.*" Thus, chemical pollution in coastal marine environments pertains to the anthropogenic release of chemical substances into coastal marine environments, which results in adverse human health, ecological and socio-economic effects.

Chemical pollution in coastal marine environments is closely linked to ocean- and land-based human activities (Williams, 1996; Islam and Tanaka, 2004; Petrosillo et al., 2009). Land-based activities include industrial and agricultural production, urban and coastal infrastructural development, tourism and mining (Simpson et al., 1996; Clark, 2001; Simboura and Zenetos, 2002; Islam and Tanaka, 2004; Petrosillo et al., 2009; Smith et al., 2009; Mestres et al., 2010; Lepland et al., 2010). Chemical pollutants from land-based activities enter coastal marine environments by various means, including river and groundwater influx, surface run-offs, atmospheric transport and deposition, direct discharges of industrial effluents and municipal outfall (Clark, 2001; Islam and Tanaka, 2004; Ruiz-Fernandez et al., 2009). Ocean-based activities such as offshore crude oil extraction, fishing, mariculture, maritime transport and dumping result in direct discharge of pollutants into the marine environment (Simpson et al., 1996; Williams, 1996; Simboura and Zenetos, 2002; Islam and Tanaka, 2004; Petrosillo et al., 2009; Simth et al., 2009; Mestres et al., 2010; Lepland et al., 2010).

With over 50% of the world's population inhabiting coastal areas (Gupta et al., 2005; Petrosillo et al., 2009), land-based activities are by far the major contributor (nearly 80%) to chemical pollution in the marine environment, with ocean-based activities (mainly maritime transport and dumping at sea) contributing about 20% (Williams, 1996). Estuaries (Meybeck and Vörösmarty, 2005; Birch et al., 2015; Alvarez-Vazquez et al., 2017) and coastal harbours (Simpson et al., 1996; Birch and Hutson, 2009; Lin et al., 2009; Smith et al., 2009; Lepland et al., 2010; Mestres et al., 2010; Schipper et al., 2010) are particularly vulnerable to chemical pollution as they are often associated with intense human activities.

Although environmental pollution can be traced back to the beginning of the history of human civilization (Islam and Tanaka, 2004; Magi and Di Carro, 2016), chemical pollution of the marine environment began to receive global attention during the mid-20th century (Magi and Di Carro, 2016), mainly as a result of global industrialisation, which resulted in an unprecedented release of chemical substances into the environment (Gaillardet et al., 2003; Wang et al., 2010). It is now known that many coastal marine environments are polluted (Islam and Tanaka, 2004; Wang et al., 2014; Vikas and Dwarakish, 2015).

2.2. Types and effects of chemical pollutants in coastal marine environments

A variety of chemical pollutants including metals, organic chemicals and radionuclides is often present in coastal marine environments from varied sources (Casado-Martinez et al., 2006; Long et al., 2006; Birch and Hutson, 2009; Schipper et al., 2010). In terms of human health and ecological significance, however, priority pollutants include (1) metals/metalloids such as Cd, Hg, Ni, Pb, Cr, Cu, Zn, Sn and As, (2) radionuclides such as ^{210}Po, ^{210}Pb, ^{226}Ra, ^{238}U, ^{232}Th, ^{228}Ra, ^{228}Th, ^{40}K and ^{137}Cs, and (3) persistent organic pollutants (POPs) such as polycyclic aromatic hydrocarbons (PAHs), polychlorinated biphenyls (PCBs), organochlorine pesticides (OCPs), and biocides such as the organotin compound tributyltin (TBT) (Williams, 1996; Volesky, 2001; Islam and Tanaka, 2004; Casado-Martinez et al., 2006; Birch and Hutson, 2009; Schipper et al., 2010; Magi and Di Carro, 2016).

TBT is notorious for imposex, a condition where females of certain species of whelk and gastropod develop male organs (Clark, 2001; Schipper et al., 2008), resulting in a decline in whelk and gastropod population (Ten Hallers-Tjabbes et al., 1994; Champ and Seligman, 1996; Mensink et al., 1996a, b; Schipper et al., 2008). These chemical pollutants are persistent and bioaccumulative with the potential to be transferred along the food chain (Chau, 2005; Nyarko et al., 2011a; Kelderman, 2012; Iqbal et al., 2013; Romano et al., 2013). They are also toxic and may exert a broad range of effects, including cancer, mutation, hormonal disruption, reproductive anomalies, death of organisms and loss of biodiversity, alteration/destruction of ecological habitats, declines in productivity, restrictions on seafood consumption, human diseases, hindrance to coastal activities, poverty, and costly remediation (Williams, 1996; Islam and Tanaka, 2004; Stoschek and Zimmermann, 2006; Lepland et al., 2010).

In recent decades, a wide range of organic compounds referred to as chemicals of emerging concern (COEC) have been recognized as potentially hazardous to human and aquatic life (Stronkhorst and van Hattum, 2003; Schipper et al., 2010; Geissen et al., 2015; Magi and Di Carro, 2016). Some COEC are halogenated organic compounds and are thus potentially persistent and bioaccumulative. They include brominated flame retardants (BFRs) such as polybrominated diphenyl ethers (PBDEs); perfluorinated chemicals (PFCs) such as perfluorooctane sulfonate (PFOS) and perfluorooctanoic acid (PFOA); polychlorinated dibenzodioxins (PCDDs), polychlorinated dibenzofurans (PCDFs) and polychlorinated naphthalenes (PCNs). Most of the halogenated COEC have been designated as priority pollutants by the European Union.

2.3. Sources of chemical pollutants in coastal marine environments

Sediment-associated pollutants in coastal marine environments may emanate from different sources. Metals may be present in sediments as a result of their natural occurrence in rocks and the Earth's crust (Clark, 2001). Anthropogenic inputs originate from e.g. combustion of fossil fuels, mining and smelting operations, industrial and manufacturing processes, and waste disposal (Zhou et al., 2008; Iqbal et al., 2013; Alvarez-Vazquez et al., 2017). They have exceeded natural or background levels in many areas (Rae, 1997; Clark, 2001; Chatterjee et al., 2007; Iqbal et al., 2013; Wang et al., 2014). Radionuclides may also be present in sediments as a result of their natural occurrence in rocks, the Earth's crust and the atmosphere as well as from anthropogenic sources. Major anthropogenic sources include the global nuclear tests conducted between the mid-1940s and the 1980s, nuclear accidents, nuclear fuel reprocessing and discharges from nuclear industries (Clifton et al., 1995; Livingston and Povinec, 2000). Anthropogenic activities such as agriculture, mining, and oil and gas development may also lead to enhancement in the environmental levels of radionuclides referred to as Technologically Enhanced Naturally-Occurring Radioactive Materials (TENORM) (UNSCEAR, 2000; Al-Trabulsy et al., 2011; Nyarko et al., 2011b).

Like metals and radionuclides, the sources of PAHs in the marine environment may be natural (e.g. natural forest fires and diagenetic processes) or anthropogenic such as waste incineration, coal processing, crude oil refining, combustion of fossil fuels and spillage of crude oil (Mai et al., 2003; Yim et al., 2007; Giuliani et al., 2008). Unlike metals, PAHs and radionuclides, OCPs have no known natural origin; they are synthetic and have been released into the environment through their use for pest control in agriculture and public health programmes (Rajendran et al., 2005; Ntow and Botwe, 2011; Thomas et al., 2012; Ahmed et al., 2015). The source of TBT in the marine environment is the use of TBT-based paints for coating of marine crafts and docks in harbours to control biofouling (Williams, 1996; Berto et al., 2007; Schipper et al., 2008; Nyarko et al., 2011a; Castro et al., 2012). Thus, commercial ports, dockyards and marinas are potential 'hot-spots' of TBT pollution. COEC have diverse sources such as pharmaceuticals, disinfection by-products, pesticides, industrial chemicals and wood preservatives (Sauvé and Desrosiers, 2014; Geissen et al., 2015).

2.4. Fate of chemical pollutants in coastal marine environments

Most chemical pollutants including radionuclides (Pfitzner et al., 2004; Yeager et al., 2005; Santschi et al., 2006; Mabit et al., 2008), metals (Santschi et al., 2001; Kelderman and Osman, 2007; Ruiz-Fernandez et al 2009; Lepland et al., 2010), OCPs and PAHs (Wang et al., 2001; Luo et al., 2004; Yang et al., 2005; Hu et al., 2009; Lin et al., 2009) that enter coastal marine environments are extremely particle-reactive. This usually results in a larger component of the pollutants partitioning to particulates (Gómez-Gutiérrez et al., 2007; Santschi et al., 2001), particularly the fine silt and clay particles, compared to the overlying water (Power and Chapman, 1992; Nyarko et al., 2011a). Thus, sediments participate in the fate of chemical pollutants in aquatic systems (Mulligan et al., 2001; Ruiz-Fernandez et al 2009; Prato et al., 2011; Jiang et al., 2013), acting as an important source of exposure of chemical pollutants to benthic organisms via direct contact or ingestion (Bat and Raffaelli, 1998; Mulligan et al., 2001; Burton, 2002; DelValls et al., 2004; Burgess et al., 2007; Gómez-Gutiérrez et al., 2007; Birch and Hutson, 2009). Metals (Rainbow, 2007; Schipper et al., 2010; Carvalho et al., 2012; Gaion et al., 2014), POPs (Burton, 2002; Islam and Tanaka, 2004; Birch and Hutson, 2009) and radionuclides (Hassona et al., 2008; Sirelkhatim et al., 2008) can subsequently accumulate in the tissues of benthic invertebrates and other organisms in the marine food web. Metals are non-degradable; radionuclides degrade through radioactive decay, whereas organic pollutants slowly degrade via weathering and biochemical conversion to other compounds (James, 2002; Rani et al., 2017). For example, the radioactive decay of ^{226}Ra and the degradation of DDT are shown in Figs. 2.1 and 2.2, respectively.

$$^{226}\text{Ra} \xrightarrow{\alpha} {}^{222}\text{Rn} \xrightarrow{\text{several } T_{1/2}} {}^{210}\text{Pb} \xrightarrow{\beta} {}^{210}\text{Bi} \xrightarrow{\beta} {}^{210}\text{Po}$$

$T_{1/2}$ 1600 y   3.82 d   22.3 y   5.01 d   138.4 d

Fig. 2.1 Radioactive decay of ^{226}Ra to ^{210}Po through several intermediate radionuclides

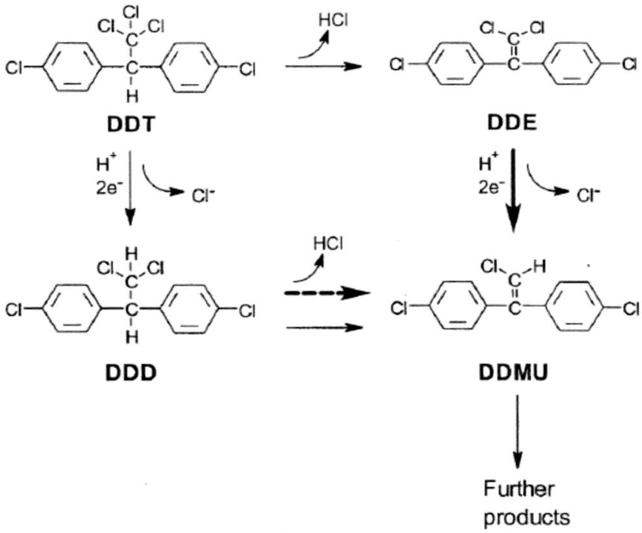

Fig. 2.2 Degradation of DDT to DDE, DDD and other metabolites (Quensen III et al., 1998).

The fate of metals (Kelderman and Osman, 2007; Di Palma and Mecozzi, 2007; Hamzeh et al., 2014), radionuclides (Sirelkhatim et al., 2008; Sugandhi et al., 2014) and organic pollutants (Lin et al., 2009) is influenced by environmental factors such as the pH and redox state. Other factors include microbiological processes (Ryding, 1985) and the presence of Fe/Mn oxides and hydroxides or organic matter (Horowitz and Elrick, 1987; Wang et al., 2001; Islam and Tanaka, 2004; Dung et al., 2013; Hamzeh et al., 2014). Sediment organic matter (OM) may derive from marine phytoplankton (autochthonous sources) or terrestrial higher plants (allochthonous sources), which may be distinguished by their total organic carbon (TOC)/total nitrogen (TN) ratios. Generally, OM of autochthonous sources is characterised by a TOC/TN ratio $\leq$ 6-8:1, while that of allochthonous sources is characterised by a TOC/TN ratio ≥ 20:1 (Burdige, 2007; Tesi et al., 2007; Guerra et al., 2010; Mahapatra et al., 2011). Thus, the TOC/TN ratios allow an assessment of the predominant source of OM in marine sediments.

2.5. Assessment of chemical pollution in coastal marine environments

In addition to their high affinity for chemical pollutants, sediments show lower variation in pollutant concentrations than their overlying water (Beiras et al., 2003). Therefore, sediments are considered as a better indicator of chemical pollution than the water phase (Sundberg et al., 2005; Denton et al., 2006; Giuliani et al., 2011; Moukhchan et al., 2013) and have become a major tool for the assessment of chemical pollution in aquatic environments (Díaz-de Alba et al., 2011; Kalwa et al., 2013).

In the GESAMP context of marine pollution, the assessment of chemical pollution in sediments should involve the 4 steps as shown in Fig. 2.3. These steps are described in detail in the text below.

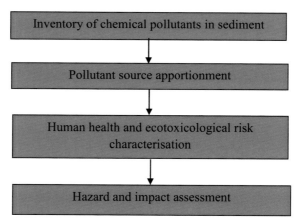

Fig. 2.3 Schematic presentation of the assessment of chemical pollution in sediments

2.5.1. Inventory of chemical pollutants in sediments

Accurate measurement of chemical pollutants is necessary in pollution assessment since it can provide information on the potentially toxic chemicals in the sediments and help identify appropriate remediation techniques. Considering that a variety of chemical pollutants may be present in marine sediments (Casado-Martinez et al., 2006; Long et al., 2006; Birch and

Hutson, 2009; Schipper et al., 2010), it is impractical to analyse the whole range of chemical pollutants in sediments. Moreover, the analysis of COEC in sediments is currently a challenge since they are typically present in very low concentrations and the development of sampling protocols and analytical techniques is at its infancy (Geissen et al., 2015). On the contrary, sampling protocols, standard analytical techniques and standard reference materials (SRMs) or certified reference materials (CRMs) that allow accurate measurements are well-developed for most of the classical chemical pollutants including radionuclides, OCPs, PAHs and metals. These pollutants are of wide interest due to their potential toxicity and adverse biological effects such as cancer, mutations and reproductive anomalies (Willett et al., 1998; Wang et al., 2001; Little, 2003; Islam and Tanaka, 2004; King et al., 2004; De Luca et al., 2004; Casado-Martinez et al., 2006; Schmid and Schrader, 2007; Hassona et al., 2008; Sirelkhatim et al., 2008; Birch and Hutson, 2009; Schipper et al., 2010; Ravanat et al., 2014).

Radionuclides of interest include ^{238}U, ^{210}Pb, ^{226}Ra, ^{232}Th, ^{228}Ra, ^{228}Th, ^{40}K and ^{137}Cs (Hassona et al., 2008; Sirelkhatim et al., 2008; Ulanovsky et al., 2008), while banned OCPs of greatest interest include DDT (dichlorodiphenyltrichloroethane) and HCH (hexachlorocyclohexane) (Walker et al. 1999; Yang et al., 2005; Hu et al., 2009; Lin et al., 2009; Singh and Lal 2009). DDT and HCH (see Fig. 2.4) are targeted for global elimination and consequently banned in many countries under the Stockholm Convention on persistent organic pollutants (Ntow and Botwe, 2011), which seeks to protect human health and the environment from the harmful effects of these pollutants. However, concerns about these banned OCPs remain due to their environmental persistence, weak enforcement of their ban, the disposal of stockpiles and the permitted use of DDT in some malaria endemic developing countries (Ntow and Botwe, 2011; Ahmed et al., 2015).

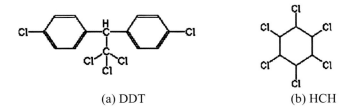

(a) DDT   (b) HCH

Fig. 2.4 Chemical structures of (a) DDT and (b) HCH

Metals of greatest concern include Pb, Cu, Zn, As, Cr, Ni, Cd, Sn and Hg (Volesky, 2001; Casado-Martinez et al., 2006; Schipper et al., 2010). PAHs are a mixture of several congeners, which may be classified as low molecular weight PAHs (containing two or three fused benzene rings) and high molecular weight PAHs (containing four to six fused benzene rings) (Yunker et al., 2000; Yunker et al., 2002; Neff et al., 2005). 16 PAHs (see Fig. 2.5) have been designated as priority pollutants by the United States Environmental Protection Agency (USEPA) and are, therefore, the main target in terms of environmental monitoring of PAHs (Nisbet and LaGoy, 1992; Wang et al 2001; De Luca et al., 2004; King et al., 2004).

2.5.2. Pollutant source apportionment in sediments

For sediment-associated chemical pollutants that may emanate from natural and anthropogenic sources, it is the anthropogenic component that is related to pollution (Chapman, 2007). Thus, Chapman and Anderson (2005) provide a distinction between "pollution" and "contamination", the former occurring when the presence of a pollutant in the environment results in harmful biological effects, while the latter refers to the presence of a pollutant that is either not normally found in the environment or above the natural background concentration. Delineating the anthropogenic sources from the natural sources is, therefore, of interest in chemical pollution assessment (Aloupi and Angelidis, 2001; Chapman, 2007; Chatterjee et al., 2007; Dung et al., 2013). Radionuclides of natural (e.g. ^{238}U, ^{210}Pb, ^{226}Ra, ^{232}Th, ^{228}Ra, ^{228}Th, ^{40}K and ^{7}Be) and artificial (e.g. ^{90}Sr, ^{137}Cs, ^{239}Pu, ^{240}Pu and ^{241}Am) sources are well known. However, natural radionuclide enrichment in sediments may occur as a result of anthropogenic activities (UNSCEAR, 2000; Al-Trabulsy et al., 2011; Nyarko et al., 2011b) and this can be assessed by comparing measured concentrations in sediments with background concentrations.

2.5.2.1.   DDT and HCH

DDT is of anthropogenic origin (Ntow and Botwe, 2011) and it undergoes degradation mainly to dichlorodiphenyldichloroethylene (DDE) and dichlorodiphenyldichlororethane (DDD) (see Fig. 2.2), which are all persistent with similar physical and chemical characteristics (WHO, 1989). DDT was commercially produced as technical DDT with a characteristic composition, which can provide insight into the timing of their use. This is relevant for assessing the effectiveness of the ban/regulation of the pesticides (Botwe et al., 2012). Technical DDT has a

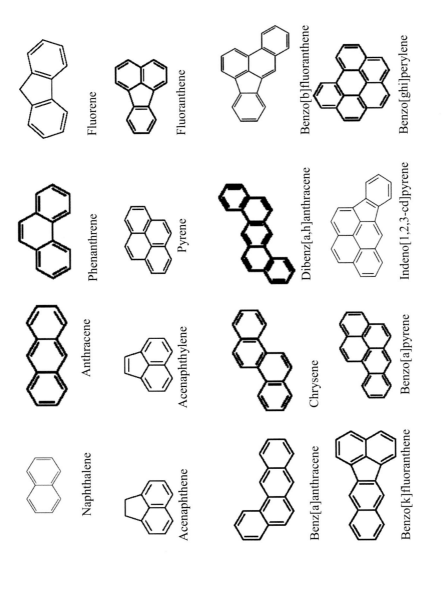

Fig. 2.5 Chemical structures of the USEPA 16 priority PAHs

composition of 77% *p,p'*-DDT, 4.0% *p,p'*-DDE and 0.3% *p,p'*-DDD. Due to the low proportions of DDE and DDD in technical DDT, the DDT/DDD and DDT/(DDE + DDD) ratios < 1.0 are generally indicative of past use of DDT, whereas ratios > 1.0 point to current use (Lin et al., 2009; Botwe et al., 2012). HCH may be present in the environment as isomers namely α-HCH, β-HCH, γ-HCH and δ-HCH (Fig. 2.6), which are all persistent with similar physical and chemical characteristics (Willett et al., 1998; Walker et al., 1999; Singh and Lal, 2009) and can undergo interconversions (Walker et al., 1999; Yang et al., 2005; Hu et al., 2009).

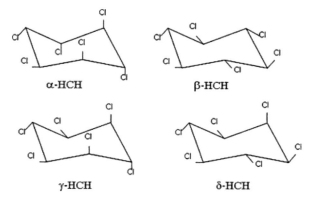

Fig. 2.6 Chemical structures of α-HCH, β-HCH, γ-HCH and δ-HCH isomers (Srivastava and Shivanandappa, 2010)

HCH was commercially produced as technical HCH and lindane. The major composition of technical HCH are α-HCH (60-70%) > γ-HCH (10-15%) > β-HCH (5-12%) > δ-HCH (6-10 %) (Willett et al. 1998; Walker et al., 1999; Singh and Lal, 2009). Interestingly, the insecticidal potency of technical HCH is mainly associated with γ-HCH despite the higher proportion of the α-HCH isomer (Willett et al., 1998; Walker et al., 1999; Singh and Lal, 2009). This knowledge provided an impetus for further refinement of technical HCH to the commercial production of high-purity γ-HCH referred to as lindane, which contains over 99% of the γ-HCH isomer with only trace amounts of the other isomers (Walker et al., 1999; Singh and Lal, 2009). The distinct HCH isomeric compositions of technical HCH and lindane provide a basis for delineation of these two HCH sources: technical HCH has α-HCH/γ-HCH ratios ranging between 3 and 7 whereas those of lindane are close to 1.0 or < 1.0 (Willett et al., 1998).

2.5.2.2. PAHs

PAH sources related to oil spills are described as petrogenic, while sources related to the incomplete combustion of organic matter and fossil fuel are described as pyrogenic (Wang et al., 2001; Yunker et al., 2002; Yan et al., 2006; Yim et al., 2007; Abrajano et al., 2007). Petrogenic sources are enriched in low molecular weight PAHs (LPAHs) but depleted in high molecular weight PAHs (HPAHs), whereas the converse holds for pyrogenic sources (Yunker et al., 2000; Yunker et al., 2002; Neff et al., 2005). The difference in the compositions of PAHs originating from petrogenic and pyrogenic sources provides a basis for the characterisation of these PAH sources in sediments: a ΣLPAH/ΣHPAH ratio > 1.0 indicates petrogenic sources, while a ratio < 1.0 indicates pyrogenic sources (Rocher et al., 2004).

Various PAH isomeric ratios such as Anthracene/(Anthracene + Phenanthrene), Fluoranthene/(Fluoranthene + Pyrene) and Benzo[a]Anthracene/(Benzo[a]Anthracene + Chrysene) (Yunker et al., 2002; Abrajano et al., 2003; Rocher et al., 2004; Nyarko et al., 2011c; Guerra, 2012) have also been used as diagnostic tools for PAH source apportionment. An Anthracene/(Anthracene + Phenanthrene) ratio < 0.10 points to a petrogenic source, while a ratio > 0.10 signifies a pyrogenic source. A Fluoranthene/(Fluoranthene + Pyrene) ratio < 0.50 is indicative of a petrogenic source, while a ratio > 0.50 indicates a pyrogenic source. A Benzo[a]Anthracene/(Benzo[a]Anthracene + Chrysene) ratio < 0.20 characterises a petrogenic source, a ratio of 0.20-0.35 indicates mixed petrogenic and pyrogenic sources, while a ratio > 0.35 points to a pyrogenic source (Yunker et al., 2000; Yunker et al., 2002).

2.5.2.3. Metals

Quantitative geochemical approaches such as the metal enrichment factor (EF) proposed by Sinex and Helz (1981) and the geo-accumulation index (I_{geo}) proposed by Muller (1969) are widely used to apportion the sources of metals in sediments. These geochemical indices compare measured concentrations of metals in sediment samples with their corresponding concentrations in pre-impacted samples to (1) evaluate metal enrichment and contamination status and (2) identify the predominant source of metals in sediments as natural or anthropogenic. Where metal concentrations in pre-impacted sediment samples are not available, metal enrichment and contamination are assessed by using metal concentrations in deep layers of sediment cores (Birch and Olmos, 2008; Abrahim and Parker, 2008; Yilgor et al., 2012), average crustal metal concentrations (Sinex and Helz, 1981; Addo et al., 2012;

Mahu et al., 2015) or average shale metal concentrations (Chatterjee et al., 2007; Addo et al. 2011, El-Sorogy et al., 2016). Metals with EFs close to 1.0 are not enriched in the sediments, are predominantly of natural origin and their associated sediments are considered "uncontaminated". On the other hand, metals with high EFs are considered enriched in the sediments, mainly of anthropogenic origin and their associated sediments are considered "contaminated".

Since metal distribution in sediments can be influenced by variations in mineralogical composition, grain size, organic matter and carbonate content of the sediments (Horowitz and Elrick, 1987; Aloupi and Angelidis, 2001; Dung et al., 2013), normalisation is required when deriving EFs in order to offset any anomaly in sediment metal distribution and allow a comparison of EFs for different sediments to be made (Sinex and Helz, 1981; Aloupi and Angelidis, 2001; Chatterjee et al., 2007; Addo et al., 2012; Yilgor et al., 2012; Dung et al., 2013; Iqbal et al., 2013; Jiang et al., 2013; Mahu et al., 2015). Although metal normalisation to the silt and clay fraction or analysis of sediments of similar grain sizes can offset the grain size effect (Horowitz, 1985; Aloupi and Angelidis, 2001; Dung et al., 2013), normalisation to a conservative metal such as Al (Aloupi and Angelidis, 2001; Chatterjee et al., 2007; Mahu et al., 2015) or Fe (Sinex and Helz, 1981; Addo et al., 2012; Yilgor et al., 2012) can offset both the grain size and mineralogical effects (Aloupi and Angelidis, 2001; Dung et al., 2013) and is, therefore, incorporated in the derivation of the EF according to Eq. 1:

$$EF = [M/N]_{Sample}/[M/N]_{Crust} \qquad (1)$$

where $[M/N]_{Sample}$ is the metal/Fe or metal/Al ratio in the sediment sample and $[M/N]_{Crust}$ is the same in the pre-impacted sediment.

Based on the EF value, the degree of metal enrichment is defined as follows: deficient (EF $\leq$ 1), minor enrichment (1 < EF $\leq$ 3), moderate enrichment (3 < EF $\leq$ 5), moderately severe enrichment (5 < EF $\leq$ 10), severe enrichment (10 < EF $\leq$ 25), very severe enrichment (25 < EF $\leq$ 50), and extremely severe enrichment (EF > 50).

The I_{geo} is derived according to Eq. (2):

$$I_{geo} = Log_2 [C_n/(1.5 \times B_n)] \qquad (2)$$

where C_n is the metal concentration in the sediment sample and B_n is the corresponding pre-impacted metal concentration.

Using the I_{geo} does not require a compensation for the mineralogical effect (Dung et al., 2013). Instead, a value of 1.5 is introduced as a background matrix correction due to lithogenic effects (Muller, 1969; Addo et al. 2011; Iqbal et al. 2013; Mahu et al. 2015). Based on the I_{geo} values, metal contamination in sediments is characterised as follows: uncontaminated ($I_{geo} <$ 0), uncontaminated to moderately contaminated ($0 \leq I_{geo} < 1$), moderately contaminated ($1 \leq I_{geo} < 2$), moderately to highly contaminated ($2 \leq I_{geo} < 3$), heavily contaminated ($3 \leq I_{geo} < 4$), highly to very highly contaminated ($4 \leq I_{geo} < 5$), and very heavily contaminated ($I_{geo} \geq 5$).

2.5.3. Human health and ecotoxicological risk characterisation

Assessment of the risk posed by contaminated sediments, i.e. the likelihood of sediment-bound contaminants to cause adverse biological effects to human and non-human biota (Birch and Hutson, 2009), is critical for the handling and management of contaminated sediments (Burton, 2002; Birch and Hutson, 2009). The risk characterisation may differ for different pollutants although in some cases, similar approaches are applicable. For radionuclides, risk characterisation approaches include the use of radiological hazard indices and the ERICA (Environmental Risk from Ionising Contaminants Assessment and Management) model. The characterisation of the potential ecotoxicological risk posed by individual metals, OCP and PAH pollutants in sediments commonly involves the use of biological effect-based numerical sediment quality guidelines. The risk assessment code is also used to characterise the potential risk of metals entering the food, while the total toxicity equivalence approach can be used to evaluate the integrated risk posed by a mixture of PAH congeners in a sediment sample.

2.5.3.1. *Radiological hazard indices and the ERICA model*

Five radiological hazard indices (RHI), namely (1) total absorbed dose rate in air (D), (2) radium equivalent activity (Ra_{eq}), (3) external hazard index (H_{ex}), (4) annual gonadal dose equivalent (AGDE), and (5) annual effective dose equivalent (AEDE) are commonly used to assess the risk posed by radionuclide contamination to human health.

The D, measured in nanogray per hour (nGy.h^{-1}), expresses the rate of human exposure to gamma radiation in air at 1m above the ground due to the activities of ^{226}Ra, ^{232}Th and ^{40}K in the sediment samples. It is calculated by applying dose conversion factors or dose coefficients for the specific activities of ^{226}Ra, ^{232}Th and ^{40}K, respectively, according to Eq. 3 (El Mamoney and Khater, 2004):

$$D \text{ (nGy.h}^{-1}) = 0.462A_{Ra} + 0.604A_{Th} + 0.0417A_K \tag{3}$$

where A_{Ra}, A_{Th} and A_K are the specific activities (Bq.kg^{-1} dry wt.) of ^{226}Ra, ^{232}Th and ^{40}K, respectively, in the sediment samples. The applied dose conversion factors for ^{226}Ra, ^{232}Th and ^{40}K are 0.462, 0.604 and 0.0417, respectively, representing their dose rates in air per unit specific activity (nGy.h^{-1}/Bq.kg^{-1}) in the sediment samples.

The Ra$_{eq}$ is a weighted sum of the specific activities of ^{226}Ra, ^{232}Th and ^{40}K in the sediment sample that allows comparison of the specific activities of different samples to be made with respect to their ^{226}Ra, ^{232}Th and ^{40}K specific activities (El Mamoney and Khater, 2004), assuming that 370 Bq.kg^{-1} of ^{226}Ra, 259 Bq.kg^{-1} of ^{232}Th and 4810 Bq.kg^{-1} of ^{40}K produce the same gamma-radiation dose rates (Kurnaz et al., 2007). It is calculated based on Eq. 4 (El Mamoney and Khater, 2004; Xinwei et al., 2006; Kurnaz et al., 2007):

$$\text{Ra}_{eq} \text{ (Bq.kg}^{-1}) = A_{Ra} + 1.43A_{Th} + 0.077A_K \tag{4}$$

where A_{Ra}, A_{Th} and A_K are the specific activities (Bq.kg^{-1} dry wt.) of ^{226}Ra, ^{232}Th and ^{40}K, respectively, in the sediment samples.

The H_{ex} is a measure of the indoor radiation dose rate associated with external gamma radiation exposure from natural radionuclides in building materials, and is important when considering the suitability of sediment for building material. It is calculated based on Eqn. 5 with the assumption of infinitely thick walls without windows or doors (Xinwei et al., 2006).

$$H_{ex} = (A_{Ra}/370) + (A_{Th}/259) + (A_K/4810) \tag{5}$$

where A_{Ra}, A_{Th} and A_K are the specific activities (Bq.kg^{-1}) of ^{226}Ra, ^{232}Th and ^{40}K, respectively, in the sediment samples. To limit the radiation exposure to the permissible dose equivalent limit of 1.5 mSv.y^{-1}, which is considered safe for humans, the value of H_{ex} must not exceed 1 (Xinwei et al., 2006; Kurnaz et al., 2007).

Owing to their relatively higher sensitivity to ionising radiation compared to other organs of the body, the gonads are considered to be at a high risk of radiation exposure and are therefore of great interest in radiological assessments (UNSCEAR, 2000; Xinwei et al., 2006; Kurnaz et al., 2007). The AGDE, measured in microsievert per year (μSv.y^{-1}), estimates the potential radiation dose that the gonads may receive from ^{226}Ra, ^{232}Th and ^{40}K. The sievert (Sv) is another unit of radiation dose from the radiological protection point of view. Unlike the Gray, it accounts for the relative biological effectiveness of different types of radiation (e.g. gamma rays, alpha and beta particles), which are known to produce dissimilar magnitudes of biological effects (Schmid and Schrader, 2007). The AGDE is calculated according to Eq. 6 (Xinwei et al., 2006; Kurnaz et al., 2007):

$$\text{AGDE } (\mu\text{Sv.y}^{-1}) = 3.09 A_{Ra} + 4.18 A_{Th} + 0.314 A_K \qquad (6)$$

where A_{Ra}, A_{Th} and A_K are the specific activities (Bq.kg^{-1}) of ^{226}Ra, ^{232}Th and ^{40}K, respectively, in the sediment samples. The constants 3.09, 4.18 and 0.314 are conversion factors expressed as μSv.y^{-1}/Bq.kg^{-1}.

The *D* can be converted to AEDE to assess the dose rate to an individual from outdoor gamma radiation over a period of one year, taking into consideration an outdoor occupancy factor of 0.2 and an effective dose conversion factor of 0.7 Sv/Gy. It is calculated from Eq. 7 (Kurnaz et al., 2007):

$$\text{AEDE } (\mu\text{Sv/y}) = D \text{ (nGy/h)} \times 8760 \text{ (h/y)} \times 0.2 \times 0.7 \text{ (Sv/Gy)} \times 10^{-3} \qquad (7)$$

For human health protection, the recommended values of *D*, Ra$_{eq}$, H_{ex}, AGDE and AEDE are 55 nGy.h^{-1}, 370 Bq.kg^{-1}, 1, 300 μSv.y^{-1} and 70 μSv/y, respectively (Xinwei et al., 2006; Kurnaz et al., 2007).

For assessment of the potential radioecological risks, the use of reliable models of wide applicability is necessary (Ulanovsky et al., 2008). One such model is the ERICA, developed by the European Commission (Larsson, 2008). The ERICA model provides an integrated approach to the assessment and management of environmental risks from ionising radiation (Beresford et al., 2007). A detailed description of the ERICA model can be found in literature (Beresford et al., 2007; Brown et al., 2008; Larsson, 2008). The ERICA model can be used to provide an overview of the potential radioactivity levels in biota and the dose rates they are likely to receive based on the measured specific activities in the sediments.

2.5.3.2. The Sediment Quality Guideline approach

Various sediment quality guidelines (SQGs) have been developed around the world (Burton, 2002) for the characterisation of the potential ecotoxicological risk posed by metal, OCP and PAH contamination in sediments. SQGs serve as national and international guidelines for the disposal of dredged materials (Mamindy-Pajany et al., 2010). Among the widely used SQGs are the effects-range low (ERL) and the effects-range median (ERM) values (Long et al., 1995; Long et al., 1998; Long et al., 2006). These SQGs were empirically derived by matching concentrations of chemical contaminants in sediments against biological effect data (Long et al., 1995; Long et al., 1998). Long et al. (1998) reported a 10% incidence of adverse biological effects in sediment-dwelling organisms when contaminant concentrations were below the ERL, but > 75% incidence of adverse biological effects when contaminant concentrations exceeded the ERM. Thus, the ERL and the ERM define the likelihood of a chemical concentration in the sediment to cause adverse biological effect with good predictive ability (Long et al., 1998).

The ERL and ERM define three concentrations ranges and their associated risks are interpreted as follows: contaminant concentration at or below the ERL indicates low risk, contaminant concentration above the ERL but below the ERM indicates moderate risk, while contaminant concentration above the ERM indicates high risk (Birch and Hutson, 2009). The use of SQGs can help identify contaminants of potential concern and areas of priority (Burton, 2002; Long et al., 2006; Birch and Hutson, 2009). Table 2.1 presents the ERL and ERM values for some priority metals, OCPs and PAHs.

It has been recognised that several chemical contaminants are often present in sediments (Casado-Martinez et al., 2006; Long et al., 2006), particularly harbour sediments (Birch and Olmos, 2008; ; Birch and Hutson, 2009; Schipper et al., 2010), the combined toxicity of which may potentially differ from those of the individual pollutants (Burton, 2002; Long et al., 2006; Birch and Olmos, 2008). To predict the ecotoxicological risks from the combined toxicity of the different pollutants present in a sediment sample, the mean ERM quotient (mERMQ) is used (Long et al., 2006; Birch and Hutson, 2009). The mERMQs are derived by normalising the concentrations of the individual contaminants to their respective ERMs, then summing all the quotients and finding their average.

Table 2.1 ERL and ERM values (mg.kg^{-1} dw) for some priority PAHs, metals and OCPs

Contaminant	ERL[*]	ERM[*]	Contaminant	ERL[*]	ERM[*]
PAHs (µg.kg^{-1})			Metals (mg.kg^{-1} dw)		
Acenaphthene	16	500	Cd	1.2	9.6
Acenaphthylene	44	640	Hg	0.15	0.71
Anthracene	85.3	1100	Ni	20.9	51.6
Fluorene	19	540	Pb	46.7	218
Naphthalene	160	2100	Cr	81	370
Phenanthrene	240	1500	Cu	34	270
Chrysene	384	2800	Zn	150	410
Fluoranthene	600	5100	As	8.2	70
Pyrene	665	2600			
Benz[a]anthracene	261	1600	OCPs (µg.kg^{-1})		
Benzo[a]pyrene	430	1600	Total DDT	1.58	46.1
Dibenz[a,h]anthracene	63.4	260	*p,p*-DDE	2.2	27

[*]Long et al. (1995); Burton (2002)

Assumptions underlying the ERMQ approach are that (1) the different pollutants contribute additively to the overall toxicity, rather than antagonistically or synergistically, and (2) samples with the same mERMQ pose similar ecotoxicological risks (Long et al., 2006; Birch and Hutson, 2009). Based on this approach, the potential ecotoxicological risk is characterised as follows: minimal if mERMQ < 0.1, low if 0.1 ≤ mERMQ < 0.5, moderate if 0.5 ≤ mERMQ < 1.5, and high if mERMQ ≥ 1.5 (Birch and Hutson, 2009).

2.5.3.3.   The total toxicity equivalence approach

PAHs are often present in sediments as complex mixtures of several congeners with wide-ranging toxic and carcinogenic potentials (Nisbet and LaGoy, 1992; Neff et al., 2005; Escher et al., 2008). In terms of toxicology, benzo[a]pyrene is the most extensively studied among the PAH congeners and considered to be highly toxic and carcinogenic (Nisbet and LaGoy,

1992; Escher et al., 2008). A toxic equivalency factor (TEF) of a PAH congener expresses its toxicity relative to that of benzo[a]pyrene, which serves as the surrogate PAH and is assigned a TEF value of 1.0 (Nisbet and LaGoy, 1992). The TEF values for the PAH$_{16}$ according to Nisbet and LaGoy (1992) are presented in Table 2.2.

Table 2.2 The USEPA 16 priority PAH congeners and their assigned TEF values

PAH congener	TEF[*]	PAH congener	TEF[*]
Acenaphthylene	0.001	Anthracene	0.01
Fluorene	0.001	Benzo[ghi]perylene	0.01
Naphthalene	0.001	Benz[a]anthracene	0.1
Phenanthrene	0.001	Benzo[b]fluoranthene	0.1
Fluoranthene	0.001	Benzo[k]fluoranthene	0.1
Pyrene	0.001	Indeno[1,2,3-cd]pyrene	0.1
Acenaphthene	0.001	Benzo[a]pyrene	1.0
Chrysene	0.01	Dibenz[a,h]anthracene	5.0

[*]Nisbet and LaGoy (1992)

For a quantitative evaluation of the integrated risk posed by a mixture of PAHs in a sediment sample, the total toxicity equivalence (TEQ) is computed from the concentrations and TEFs of the individual PAH congeners according to Eq. 8 (Escher et al., 2008; Nyarko et al., 2011c):

$$TEQ \ (mg.kg^{-1}) = \Sigma(Cn \ x \ TEF_n) \hspace{3cm} (8)$$

where Cn and TEF$_n$ are, respectively, the concentration (mg.kg^{-1}) and TEF of an individual PAH in the sediment sample.

The underlying assumptions of the TEQ approach are that the PAH congeners in the mixture (1) exhibit the same mode of action and (2) contribute additively to the overall toxicity (Escher et al., 2008). Based on the TEQ approach, sediment contamination is classified as follows (Yang et al., 2014): uncontaminated (with potential no risk) if TEQ < 0.1 mg.kg^{-1}; slightly contaminated (with potential medium risk) if 0.1 < TEQ < 1 mg.kg^{-1}; and significantly contaminated (with potential high risk) if TEQ ≥ 1.0 mg.kg^{-1}.

2.5.3.4. *Using the risk assessment code*

Metals exist in different binding forms with different binding strengths, i.e. water-soluble and exchangeable (bound to carbonates), reducible (bound to iron and manganese oxides) and oxidisable (bound to sulphides or organic matter) forms (Díaz-de Alba et al., 2011; Kalwa et al., 2013; Kelderman and Osman, 2007). These metal binding forms are associated with different mobility, bioavailability and thus ecotoxicological potential (Calmano et al., 1993; Jain, 2004; van Hullebusch et al., 2005; Dung et al., 2013; Kalwa et al., 2013; Pini et al., 2015). Metals in the water-soluble and exchangeable form are the most weakly bound and, therefore, pose the greatest risk of entering the food chain (Calmano et al., 1993; DelValls et al., 2004; Jain, 2004; Kelderman and Osman, 2007; Dung et al., 2013; Iqbal et al., 2013; Kalwa et al., 2013). The risk assessment code (RAC) has been developed to estimate the potential risk of a sediment-bound metal entering the food chain based on the metal fractionation in sediments (Jain, 2004; Zhu et al., 2012; Jiang et al., 2013). The RAC of a given sediment-bound metal is expressed as the percentage of its concentration in the water-soluble and exchangeable form according to Eq. 9:

$$RAC = [M_{WSE}]/[M_T] \times 100\% \qquad\qquad (9)$$

where $[M_{WSE}]$ is the concentration of a given sediment-bound metal in the water-soluble and exchangeable form and $[M_T]$ is its total concentration.

Based on the value of the RAC, risk may be defined as follows (Jain, 2004; Zhu et al., 2012): <1 % indicates no risk, 1-10 % indicates low risk, 11-30 % indicates medium risk, 31-50% indicates high risk, and >50 % indicates very high risk.

A widely used analytical technique for metal fractionation in sediments is the 3-step sequential extraction scheme developed by the European Standards, Measurements and Testing Programme, formerly known as the Community Bureau of Reference (BCR) (Quevauviller et al., 1994; Quevauviller et al., 1997; Rauret et al., 1999; Ptistišek et al., 2001; Pueyo et al., 2001; Davidson et al., 2006). A major advantage of the 3-step BCR sequential extraction method over other analytical techniques such as the 5-step sequential extraction method by Tessier et al. (1979) or its modified versions (e.g. Kelderman and Osman, 2007) is that it has been standardised by inter-laboratory comparison and the use of certified reference materials (Quevauviller et al., 1994; Quevauviller et al., 1997; Rauret et al., 1999). Thus, the BCR method can ensure greater quality assurance and control.

2.5.3.5. Limitations of the RHI, SQG, TEQ and RAC approaches

The RHI, SQG, TEQ and RAC approaches constitute a screening-level ecotoxicological risk assessment, intended to predict the adverse biological effects of measured contaminants. Screening-level ecotoxicological risk assessment provides a first line of evidence of (1) the potential of measured contaminants to cause adverse biological effects, (2) contaminants of potential concern, and (3) priority areas of concern (Long et al., 1998; Burton, 2002; Long et al., 2006; Mamindy-Pajany et al., 2010), but is inadequate for a comprehensive assessment of sediment pollution (Mamindy-Pajany et al., 2010). The RHI, SQG, TEQ and RAC approaches may over or under estimate the potential hazard and impact of sediment-bound contaminants, since they do not consider:

(1) Contaminant uptake and regulation (Rainbow, 2007; Rainbow and Luoma, 2011). The RAC, for example, may not give a true reflection of metal bioavailability, which is difficult to measure (DelValls et al., 2004) due to the potential for metal regulation by organisms following uptake (Marsden and Rainbow, 2004; Rainbow, 2007; Rainbow and Luoma, 2011).

(2) Synergistic or antagonistic effects of multiple contaminants on their overall toxicity (Ciarelli et al., 1998; Burton, 2002; Simboura and Zenetos, 2002; Eggen et al., 2004; Long et al., 2006).

(3) Chronic effects that may result from exposure to low contaminant concentration over a long time and multiple effects that may be induced by a single contaminant (Eggen et al., 2004).

(4) Sediment contaminants that have not been characterised for their toxicity (Schipper et al., 2010).

(5) The entire spectrum of potentially toxic contaminants that is present in the sediment as it is difficult to measure (Escher et al., 2008) since their measurement is not practical, and SQGs have not yet been developed for all potential pollutants in sediments including the COEC (Long et al., 1995; Burton, 2002; Long et al., 2006; Schipper et al., 2010).

2.5.4. Hazard/Toxicity and impact assessment

Hazard/toxicity and impact assessment is intended to evaluate biological responses to chemical toxicity in sediments. Whole-sediment toxicity bioassays, which involve exposure of relevant organisms to contaminated sediments under controlled conditions and measurement of biological responses/endpoints (Ciarelli et al., 1998; Forrester et al., 2003; DelValls et al., 2004), are an important tool in hazard and impact assessment of contaminated sediments (DelValls et al., 2004; Annicchiarico et al., 2007; Casado-Martinez et al., 2007; Escher et al., 2008; Morales-Caselles et al., 2009; Ré et al., 2009; Schipper et al., 2010; Prato et al., 2015). The major advantage of whole-sediment toxicity bioassays over the RHI, SQG, TEQ and RAC approaches is that they integrate the toxic effects of all contaminants present in a sediment sample (Forrester et al., 2003; Escher et al., 2008) and provide a further line of evidence of the potential ecotoxicological effects of contaminated sediments (Long et al., 2006; Morales-Caselles et al., 2008; Schipper et al., 2008; Morales-Caselles et al., 2009). Thus, whole-sediment toxicity bioassays are now required in many developed countries such as Spain, Belgium, Germany (DelValls et al., 2004; Casado-Martinez et al., 2007), the Netherlands (Casado-Martinez et al., 2006), Australia (Rose et al., 2006; van Dam et al., 2008) and Italy (Prato et al., 2011) for adequate assessment of dredged sediments to support the licensing of their subsequent disposal.

For a successful whole-sediment toxicity bioassay, there is a need to identify suitable test organisms with measureable biological responses/endpoints that can discriminate between different levels of chemical toxicity in sediments (Ciarelli et al., 1998). These endpoints include acute effects such as mortality, sub-lethal and/or chronic effects such as growth and reproduction or bioaccumulation (Forrester et al., 2003; DelValls et al., 2004; Annicchiarico et al., 2007; Morales-Caselles et al., 2008; Schipper et al., 2008). The bioaccumulation assessment is intended to give an indication of the bioavailability of sediment-associated chemical contaminants and their potential for transfer along the food chain (DelValls et al., 2004). The criteria for the selection of test organisms include their importance and abundance in the ecosystem under investigation (Connon et al., 2012). For estuarine and marine whole-sediment toxicity bioassays, algae, molluscs, sea urchins, amphipods and polychaetes have been widely used as test organisms. Commercial acute sediment toxicity bioassay kits such as Microtox®, which is based on the natural bioluminescence inhibition of the marine bacteria *V.*

fischeri, are also used (Morales-Caselles et al., 2007; Libralato et al 2008; Morales-Caselles et al., 2009). Table 2.3 presents examples of sediment bioassay tests in use around the world.

Table 2.3 Test organisms used in sediment bioassay tests around the world

Sediment bioassay test organisms	References
Bacteria (Microtox®)	Beg and Ali, 2008; Libralato et al., 2008; Morales-Caselles et al., 2007, 2009; Ghirardini et al., 2009; Baran and Tarnawski, 2013
Amphipod	Casado-Martinez et al., 2007; Scarlett et al., 2007; van den Heuvel-Greve et al., 2007; Morales-Caselles et al., 2007; Mayor et al., 2008; Ré et al., 2009; de-la-Ossa-Carretero et al., 2012; Hanna et al., 2013.
Algae	Oehlmann, 2002; Mariño-Balsa et al., 2003; Chen et al., 2009
Sea urchin	Stronkhorst et al., 1999; Schipper et al., 2008; Pagano et al., 2017a, b.
Molluscs	Oehlmann, 2002; Mariño-Balsa et al., 2003
Polychaetes	Thain and Bifield, 2002; Casado-Martínez et al., 2006; Moreira et al., 2006; Mayor et al., 2008; Kalman et al., 2012

Among the various test organisms, the amphipod *Corophium volutator* (Stronkhorst et al., 2003; Casado-Martinez et al., 2007; Scarlett et al., 2007; van den Heuvel-Greve et al., 2007; Morales-Caselles et al., 2007; Mayor et al., 2008) and the polychaete *Hediste diversicolor* (Casado-Martínez et al., 2006; Moreira et al., 2006; Mayor et al., 2008) have gained popularity. The popularity of *C. volutator* (Fig. 2.7a) is mainly due to the following reasons (Ciarelli et al., 1998; Roddie and Thain, 2002; Scaps, 2002; Bat, 2005): (1) it is available in the field throughout the year, (2) it is easy to collect and maintain under laboratory conditions, (3) it shows tolerance to a wide range of salinities, sediment grain sizes and organic carbon contents, and (4) a standard protocol has been developed using *C. volutator* (Roddie and Thain 2002; Schipper et al., 2006). *H. diversicolor* (Fig. 2.7b) is also preferred as a test organism mainly due to the following attributes (Scaps, 2002; Philippe et al., 2008): (1) it commonly occurs in intertidal areas, (2) it is able to survive in hypoxic and contaminated environments, (3) it is tolerant to wide fluctuations in salinity and temperature, and (4) a standard protocol has been developed using *H. diversicolor* (Hannewijk et al., 2004). Furthermore, both *C. volutator* and *H. diversicolor* have wide geographic distributions and can be found in polar, temperate and tropical marine regions (Bat, 2005; Moreira et al., 2006; Uwadiae, 2010; Carvalho et al., 2012). However, whole-sediment toxicity bioassays with tropical species are not yet well developed (Adams and Stauber, 2008).

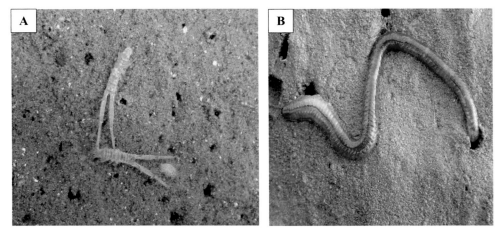

Fig. 2.7 (A) *Corophium volutator* (http://www.aquatonics.com/Corophium.jpg); (B) *Hediste diversicolor* (https://www.flickr.com/photos/gwylan/2170368562/) used as test organisms in Chapter 6.

By using whole-sediment toxicity bioassays, the potential hazard and impacts are evaluated by comparing measured biological responses with corresponding responses using uncontaminated (control/reference) sediments under similar conditions (Stronkhorst et al., 2003; DelValls et al., 2004; Mamindy-Pajany et al 2010) and applying established guidelines (Thain and Bifield, 2002; Roddie and Thain, 2002; Casado-Martinez et al., 2007; ICES, 2008). Whole-sediment toxicity bioassays can help identify contaminated areas (Ciarelli et al., 1998), but they cannot provide information on the contaminants responsible for sediment toxicity (Escher et al., 2008). This can be achieved by integrating whole-sediment toxicity bioassays with chemical analysis (Annicchiarico et al., 2007; Casado-Martinez et al., 2007; Morales-Caselles et al., 2007, 2008; Escher et al., 2008; Mamindy-Pajany et al., 2010; Prato et al., 2011). This integrated approach is required to better characterise the hazard, risk and impact of contaminated sediments (Morales-Caselles et al., 2008; Schipper et al., 2008; Mamindy-Pajany et al., 2010) and are incorporated into regulatory programmes in many developed countries (Ciarelli et al., 1998). Moreover, the use of a battery of whole-sediment bioassays (DelValls et al., 2004; Morales-Caselles et al., 2009) with different test species having potentially different sensitivities to chemical toxicity is recommended for the purpose of managing dredged materials (Annicchiarico et al., 2007; Casado-Martinez et al., 2007; Escher et al., 2008; Ré et al., 2009; Schipper et al., 2010; Prato et al., 2011; Prato et al., 2015).

2.6. Sedimentation in harbours

2.6.1. Sediment fluxes in harbours

Sediments play an important role in the ecological functioning of aquatic ecosystems, providing food for filter-feeding invertebrates (Schipper et al., 2008; Jiang et al., 2013) and spawning areas for many organisms (Prato et al., 2011). However, high influx of sediments can cause ecological problems such as smothering of benthic eggs and larvae, reduced light availability, and release of associated pollutants into the water column (Green and Coco, 2014). Sediment influx into coastal harbours is driven by natural factors such as tides, waves (Leys and Mulligan, 2011), riverine sediment discharge (Akrasi, 2011) coupled to anthropogenic factors such as sand winning (Kusimi and Dika, 2012).

The concentrations and fluxes of sediments in the water column can also influence the dynamics and transport of particle-reactive pollutants in aquatic ecosystems (James, 2002; Syvitski, et al., 2005; Lepland et al., 2010; Luo et al., 2010; George et al., 2012; Souza and Lane, 2013). Thus, sedimentation provides a means of pollutant transport to and subsequent accumulation in bottom sediments. In depositional environments, the continuous deposition of sediments without an interruption in the sedimentary sequence can serve as an archive of environmental changes (Kannan et al., 2005; Giuliani et al., 2008; Giuliani et al., 2015) and dating of sediment cores can provide a historical record of SARs and pollution events (Giuliani et al., 2008; Tang et al., 2008; Díaz-Asencio et al., 2009; Smith et al., 2009; Lepland et al., 2010; Bellucci et al., 2012).

Coastal harbours tend to have low-energy hydrodynamics and restricted water movement due to their breakwaters (Mestres et al., 2010). This facilitates the deposition and subsequent accumulation of sediments within the harbour basins (Lepland et al., 2010; Luo et al., 2010).

Sediment accumulation within harbour basins is a major issue as it results in siltation and poses navigational problems (Syvitski, et al., 2005; Van Rijn, 2005; Green and Coco, 2014). Many harbours around the world have been affected by siltation, resulting in costly dredging of several million tonnes of sediments to maintain appropriate water depths for safe navigation and/or removal of contaminated sediments (Qu and Kelderman, 2001; Barneveld and Hugtenburg, 2008; Schipper et al., 2010). The assessment of settling fluxes (SFs) and SARs can provide complementary information for understanding sedimentation processes in harbours (Liu et al 2014), which is necessary for harbour management (Buesseler et al., 2007; Leys and Mulligan, 2011).

2.6.2. Quantification of SFs

Sediment traps in the form of close-ended cylindrical tubes are important tools for the collection and assessment of settling particulate matter (and its associated pollutants) falling vertically through the water column to the bottom sediments (Hakanson, 2006; Buesseler et al., 2007; de Vicente et al., 2010; Kelderman et al., 2012; Santos-Echeandía et al., 2012; Liu et al., 2014; Szmytkiewicz and Zalewska, 2014; Helali et al., 2016). The quantity of particulate matter collected by the traps depends on the deployment period and the efficiency of trap collection (Hakanson, 2006). Sediment traps are deployed typically over a period of two weeks to allow sufficient particulate matter to be collected for analysis, while ensuring minimal organic matter decomposition (Hakanson and Jansson, 1983; Buesseler et al., 2007; Kelderman et al., 2012). Trap efficiency can be optimised through appropriate trap design. Sediment traps with aspect ratios (i.e. height/diameter) greater than 5 minimise current-induced resuspension within the traps and enhance the efficiency of traps to collect and accumulate settling particulate matter (Bloesch and Burns, 1980; Kelderman et al., 2012; de Vicente et al., 2010). The sediment trap-derived SF is expressed as the mass of dry material deposited per unit area per unit time, i.e. $mg.m^{-2}.d^{-1}$ or $g.m^{-2}.y^{-1}$. A schematic presentation of sediment trap deployment in Chapters 7 and 8 is shown in Fig. 2.8.

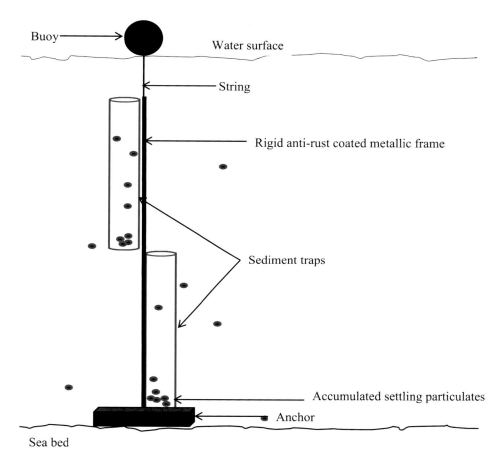

Fig. 2.8 Schematic presentation of sediment trap deployment used as samplers for settling particulate matter in Chapters 7 and 8

2.6.3. Quantification of SARs

In harbours, reservoirs, estuaries and coastal areas, bathymetric data such as tidal levels, water depth and positioning acquired at different periods can be used to quantify SARs with the application of a series of corrections (Khaba and Griffiths, 2017; Brucker et al., 2007). For shallow waters with less than 20 m water depth such as harbours, bathymetric data can be best obtained by using Phase Differencing Bathymetric Sonar (PDBS) systems owing to their associated total vertical uncertainty of 0.26 m for a 10 m water depth (Brisson et al., 2014). This uncertainty is acceptable for water depth control purposes, but it is too coarse for estimating SARs. Therefore, PDBS systems are usually suitable for mapping SARs in areas where accretion rates are high, exceeding 0.3 m.y^{-1} (Brucker et al., 2007) with time lapses of the order of decades (Ortt et al., 2000). This method provides mean SAR in the time lapse and cannot provide information on processes affecting the depth distribution of chemical pollutants. In this regard, particle-reactive radionuclides such as ^{210}Pb have found useful applications in the study of sedimentary processes and quantification of SARs (Erten, 1997; Caroll and Lerche, 2003; Giffin and Corbett, 2003; Corbett et al., 2009). Non-radioactive dating methods based on e.g. fossil markers and pollen can provide important stratigraphic time-markers, but in contrast to the radiometric dating methods, these are of limited applicability and cannot determine absolute chronologies of sediments (Carroll and Lerche, 2003; Abril, 2015).

2.6.3.1. ^{210}Pb as a tool for the quantification of SARs

The ^{210}Pb radioisotope, based on its short half-life of 22.3 years, is particularly suitable for chronological assessment and quantification of recent sedimentation rates on time-scales spanning the past 100-150 years (Lu, 2007; Appleby, 2008; Díaz-Asencio et al., 2009). Alpha spectrometry and low background gamma spectrometry are the most common methods of analysing the specific activity of ^{210}Pb in sediments (Sikorski and Bluszcz, 2008; de Vleeschouwer et al., 2010; Mabit et al., 2014) with comparable results (Zaborska et al., 2007; Sikorski and Bluszcz, 2008).

The ^{210}Pb radioisotope is a member of the ^{238}U-decay series and may be produced from the decay of ^{222}Rn in the atmosphere referred to as unsupported/excess ^{210}Pb or ^{210}Pb$_{exc}$ (Erten, 1997; Alonso-Hernandez et al., 2006; Lu, 2007). The *in situ* radioactive decay of ^{226}Ra in the water column also produces ^{210}Pb referred to as supported ^{210}Pb or ^{210}Pb$_{supp}$ (Oldfield and Appleby, 1984; Lu, 2007; Appleby, 2008; Nehyba et al., 2011; Abril and Brunskill, 2014). The ^{210}Pb$_{supp}$ is assumed to be in secular equilibrium with ^{226}Ra and other members of the decay chain preceding it, while the ^{210}Pb$_{exc}$ represents that part of the measured total ^{210}Pb specific activity in sediment that exceeds the specific activity in secular equilibrium with ^{226}Ra (Pfitzner et al., 2004; Hollins et al., 2011). The fluxes of ^{210}Pb in sediments are linked to wet and dry atmospheric deposition (Erten, 1997; Appleby, 2008; Pfitzner et al., 2004; Mabit et al., 2014) as well as surface run-off and erosion from the catchment (Mabit et al., 2008; Ruiz-Fernández et al., 2009). The sources and pathways of ^{210}Pb inputs in aquatic systems are shown in Fig. 2.9.

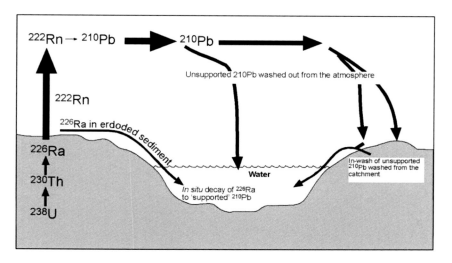

Fig. 2.9 Sources and pathways of ^{210}Pb inputs in aquatic systems
(http://www.ozcoasts.gov.au/glossary/images/pb210_diagram.jpg)

In aquatic systems, ^{210}Pb$_{exc}$ is scavenged by particulate matter, particularly the silt-clay fraction of sediments (Pfitzner et al., 2004), and subsequently accumulates in bottom sediments (Erten, 1997). The optimal material for ^{210}Pb dating is undisturbed sediments (Santschi et al., 2001; Bellucci et al., 2012). For ^{210}Pb dating of sediment cores, it is assumed that the ^{210}Pb$_{supp}$ is at equal specific activity with its parent ^{226}Ra nuclide and is constant down the sediment core profile, while the ^{210}Pb$_{exc}$ decays because it is not supported by ^{226}Ra via ^{222}Rn (Erten, 1997). The ^{210}Pb$_{exc}$ specific activity in the sediment core is obtained as the difference between the measured total ^{210}Pb specific activity and the ^{210}Pb specific activity in equilibrium with ^{226}Ra (Erten, 1997), which usually occurs in the deep layers (Appleby and Oldfield, 1978). ^{210}Pb$_{exc}$ decays in accordance with the radioactive decay law (Lu, 2007; Appleby, 2008) and can be described by Eq. (10):

$$A = A_0 e^{-\lambda t} \tag{10}$$

where A_0 is the initial specific activity of ^{210}Pb$_{exc}$, A is the specific activity of ^{210}Pb$_{exc}$ after time t, and λ is the decay constant of ^{210}Pb (0.03114 y^{-1}).

An undisturbed sedimentary sequence is thus characterised by an exponential decrease in $^{210}Pb_{exc}$ activity down the sediment core (see Fig. 2.10); the vertical $^{210}Pb_{exc}$ profile being a function of the sediment accumulation rate (Alonso-Hernandez et al., 2006; Appleby, 2008).

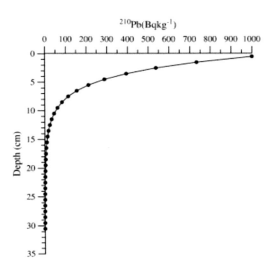

Fig. 2.10 $^{210}Pb_{exc}$ activity profile in an undisturbed sediment core (Mackenzie et al., 2011)

2.6.3.2. ^{210}Pb-based sediment dating models

Three widely used ^{210}Pb sediment dating models are the Constant Flux Constant Sedimentation (CF-CS), Constant Rate of Supply (CRS) and Constant Initial Concentration (CIC) models (McDonald and Urban, 2007; Appleby, 2008). The CIC and CRS models (Appleby and Oldfield, 1978; Robbins, 1978) are, however, considered as the two standard approaches (Appleby, 2004; Appleby, 2008; Nehyba et al., 2011). These ^{210}Pb sediment dating models convert the $^{210}Pb_{exc}$ specific activities in sediment layers into numerical ages to obtain an age-depth profile of sediment chronologies (Gunten et al., 2009).

2.6.3.2.1. The Constant Flux Constant Sedimentation (CF-CS) model

The CFCS model is a common particular case of CIC and CRS models. The CF-CS model applies under the following assumptions: (1) a constant flux of $^{210}Pb_{exc}$ from the atmosphere and a constant dry-mass sedimentation rate (Appleby and Oldfield, 1983; Díaz-Asencio et al., 2009) and (2) no post-depositional sediment mixing with every sediment layer having the same initial ^{210}Pb$_{exc}$ specific activity (Appleby and Oldfield, 1983). Under these situations, the ^{210}Pb$_{exc}$ specific activity varies exponentially in accumulated sediments (Appleby and Oldfield, 1983; Appleby, 2004; McDonald and Urban, 2007; Nehyba et al., 2011). Due to potential effects of sediment compaction, the ^{210}Pb$_{exc}$ specific activity is related to the mass-depth (g.cm^{-2}) instead of the linear-depth of the sediment core (Erten, 1997; Lu, 2007). Thus, under the CF-CS model, the expected specific activity at a given mass-depth, m, is given by Eq. 11 (Erten 1997; McDonald and Urban, 2007):

$$A_x = A_0\, e^{-\lambda m/w} \qquad\qquad\qquad (11)$$

where A_x (Bq.kg^{-1}) is the ^{210}Pb$_{exc}$ specific activity at mass-depth x; A_0 (Bq.kg^{-1}) is the initial ^{210}Pb$_{exc}$ specific activity at $x = 0$; m is the cumulative dry-mass (g.cm^{-2}), λ is the decay constant for ^{210}Pb (0.03114) and w is the mass sedimentation rate (g.cm^{-2}.y^{-1}).

A logarithmic plot of the ^{210}Pb$_{exc}$ specific activity versus mass depth will generate a linear profile with slope $= - \lambda/w$, from which the mass sedimentation rate, w, can be calculated (e.g. see Fig. 2.11).

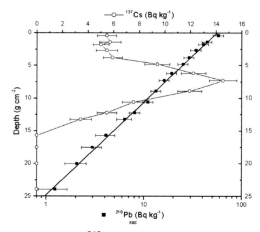

Fig. 2.11 A logarithmic plot of the ^{210}Pb$_{exc}$ specific activity versus mass depth (Díaz-Asencio et al. 2009)

2.6.3.2.2. The Constant Initial Concentration (CIC) model

The main assumption of the CIC model (Robbins, 1978; Appleby and Oldfield, 1978) is that the initial $^{210}Pb_{exc}$ specific activity (A_0) at the sediment-water interface remains constant over time with a constant sedimentation (Appleby, 2008; Nehyba et al., 2011), but allows co-variation of both $^{210}Pb_{exc}$ flux and sedimentation (Appleby, 2008). If these assumptions apply, A_0 will decline exponentially with age so that the $^{210}Pb_{exc}$ specific activity in a sediment layer of age t (A_t) is given by Eq. 12:

$$A_t = A_0 e^{-\lambda t} \qquad (12)$$

where A_0 is the $^{210}Pb_{exc}$ specific activity (Bq.kg^{-1}) at time zero, which in the absence of surface mixing represents the concentration at the sediment-water interface; A_t is the $^{210}Pb_{exc}$ specific activity (Bq.kg^{-1}) after time t and λ is the radioactive decay constant for ^{210}Pb (0.03114 y^{-1}).

Eq. 12 can be rewritten as Eq. 13 and used to calculate the age (t) of a sediment layer at a depth x (Appleby and Oldfield, 1978):

$$t = 1/\lambda \ \ln[A_0/A_x] \qquad (13)$$

If it is assumed that the sedimentation rate (S) is constant within the sediment layer x, then the period over which the sediment has accumulated is $t = x/S$. Substituting $t = x/S$ into Eq. 13 and re-arranging Eq. 13 gives Eq. 14:

$$(1/x) \ln [A_x/A_0] = - \lambda/S \qquad (14)$$

A linear regression profile of $\ln(A_x)$ versus depth (x) is expected to produce a slope $= - \lambda/S$, from which S can be computed according to Eq. 15:

$$S = - \lambda / \text{slope} \qquad (15)$$

Application of the CIC model seems difficult within an industrial harbour such as the Tema Harbour, considering the changing industrial activities and harbour configuration, which could lead to varying proportions of sediment constituents and then to varying initial concentrations of unsupported ^{210}Pb.

2.6.3.2.3. The Constant Rate of Supply (CRS) Model

The CRS model (Appleby and Oldfield, 1978; Robbins, 1978) assumes: (1) a constant rate of $^{210}Pb_{exc}$ supply to the sediments despite variations in the dry-mass sedimentation rate and (2) no post-depositional mixing (McDonald and Urban, 2007). Thus, the CRS model allows changes in sedimentation rates while estimating sediment ages (Appleby, 2008; Nehyba et al., 2011; de Souza et al., 2012). Sediment ages are derived from the fraction of the depth-integrated $^{210}Pb_{exc}$ present above the core depth considered. Thus, any deviation from an exponential decrease with depth of $^{210}Pb_{exc}$ is interpreted by the CRS model to reflect a variation of the sedimentation rate (Appleby, 2004; MacKenzie et al., 2011; Nehyba et al., 2011). Important considerations for use of the CRS model are (1) an inverse relationship between the initial concentration and the sedimentation rate, (2) a non-monotonic decrease in ^{210}Pb activity and (3) the ^{210}Pb dating horizon (i.e. the depth where the $^{210}Pb_{exc}$ specific activity is diminished and $^{210}Pb_{supp}$ specific activity becomes constant) is reached (Appleby, 2008). Using the CRS model, the age (t) of sediments at a depth x can be estimated from Eq. 16 (Appleby and Oldfield, 1978; Erten, 1997; MacKenzie et al., 2011; de Souza et al., 2012):

$$t = (1/\lambda) \ln[A_\infty/A_x] \qquad (16)$$

where A_∞ is the integrated $^{210}Pb_{exc}$ ($^{210}Pb_{exc}$ inventory) of the entire sediment column and A_x is the integrated $^{210}Pb_{exc}$ below a depth x of the sediment core.

The dry mass sedimentation rate (w, with a unit of $g.cm^{-2}.y^{-1}$) can be obtained for each two adjacent layers from Eq. 17 (Appleby and Oldfield, 1978):

$$w = \lambda A_x/C \qquad (17)$$

where C is the $^{210}Pb_{exc}$ specific activity at the depth x.

If a steady state supply of $^{210}Pb_{exc}$ to the sediment is assumed, then the inventory of $^{210}Pb_{exc}$ can be estimated using Eq. 18:

$$I_0 = \Sigma(A_i \, \rho_i \, h_i) \qquad (18)$$

where I_0 is the $^{210}Pb_{exc}$ inventory in the sediment (Bq.cm^{-2}); A_i is the $^{210}Pb_{exc}$ specific activity for sediment layer i (Bq.g^{-1}), ρ_i is the bulk density of sediment layer i (g.cm^{-3}) and h_i is the thickness of sediment layer i (cm).

The flux of $^{210}Pb_{exc}$ to the sediment (F) can be estimated from Eq. 19:

$$F = \lambda\, I_0 \tag{19}$$

2.6.3.3.   The Time Estimates from Random Entries of Sediments and Activities (TERESA) model

It has been shown that the assumption of an ideal deposition, i.e. the deposition of new radioactive inputs above the previously existing material at the sediment-water interface (SWI), of the conventional ^{210}Pb dating models is unrealistic, particularly in sediment cores with very high porosities (Abril and Gharbi, 2012). Thus, using statistical analysis of a database of laminated sediment cores, Abril and Brunskill (2014) reconstructed historical records of initial $^{210}Pb_{exc}$ specific activity, $^{210}Pb_{exc}$ flux onto the SWI and SARs, which revealed wide temporal fluctuations. Their results also showed a linear relationship between the $^{210}Pb_{exc}$ fluxes and SAR. However, no statistically significant correlation was found between the initial $^{210}Pb_{exc}$ specific activity and SAR, which conflicts with the assumption of most ^{210}Pb-based radiometric dating models. This paved the way for the development of the TERESA model (Abril, 2016).

The TERESA model is based on the findings that (1) the flux of $^{210}Pb_{exc}$ to the SWI varies with time and (2) the flux of $^{210}Pb_{exc}$ is governed by the flux of matter instead of direct atmospheric $^{210}Pb_{exc}$ deposition, resulting in a statistically significant correlation between the flux of $^{210}Pb_{exc}$ and the flux of matter/sediment accumulation rates (Abril and Brunskill, 2014). Other important assumptions in the application of the TERESA model are that (1) $^{210}Pb_{exc}$ is a particle-bound tracer, which is continually deposited onto the SWI, (2) no post-depositional redistribution occurs and (3) the sedimentary sequence is continuous with none of its layers missing.

The TERESA model works with continuous probability distributions for initial $^{210}Pb_{exc}$ specific activities (A_0) and SAR and, therefore, assumes there are neither flood layers nor other instantaneous inputs to the sediment column (Abril, 2016). Thus, a sediment core cut into N slices (labelled $i = 1, 2, ... N$) is assumed to have resulted from the accumulated mass flow onto the SWI over the time interval ΔT_i with an average mass flow (SAR) value of w_i (Abril, 2016). If $\overline{w}$ is the magnitude of the arithmetic mean of all the w_i for the N slices, then the values of w_i are almost normally distributed around $\overline{w}$ with a standard deviation σ_w (Abril and Brunskill, 2014). This results in a linear correlation between the $^{210}Pb_{exc}$ fluxes and SAR. Thus, each sediment slice can be characterised by a pair of values ($A_{0,i}$, w_i), the sequence of which produces the $^{210}Pb_{exc}$ vertical profile, the chronology and the particular SAR history (Abril, 2016).

It is worthy to note that for the TERESA model, the mass depth m (with physical dimensions $M.L^{-2}$) is a more appropriate magnitude than the linear depth (with physical dimension L) due to (1) compaction during the sediment accretion and (2) core shortening during the coring operation and later handling (Abril, 2016). Moreover, each sediment slice of mass thickness Δm_i should be provided with its respective uncertainty, derived from the experimentally averaged specific $^{210}Pb_{exc}$ specific activities $A_i(m)$ (Abril, 2016).

Abril (2016) has validated the TERESA model using synthetic and real data from laminated sediment cores from the Santa Barbara Basin (Koide et al., 1973) and the Sihailongwan Lake, China (Schettler et al., 2006a, 2006b), for which independent chronology has been established. Prior to this thesis, the potential of the TERESA model for dating recent sediments from environments where the conventional ^{210}Pb-based dating models face constraints such as coastal harbours had not been explored.

2.6.3.4. Validation of ^{210}Pb dating models

For any of the ^{210}Pb dating models, a validation of the ^{210}Pb-based sediment chronology by other time markers is necessary (Smith, 2001; Abril, 2004; Appleby, 2004) and errors associated with the estimated sediment ages and sedimentation rates should be quantified. The short-lived ^{137}Cs isotope ($T_{1/2} = 30.1$ y) has been commonly used to validate ^{210}Pb dates of sediment cores (Erten 1997; Smith, 2001; Appleby, 2002; Pfitzner et al., 2004; Alonso-Hernandez et al., 2006; McDonald and Urban, 2007; Díaz-Asencio et al., 2009; Smith et al., 2009; Lepland et al., 2010; Giuliani et al., 2011), which can be measured directly by gamma spectrometry (Bellucci et al., 2012; Díaz-Asencio et al., 2009; Lepland et al., 2010; Smith et al., 2009). The main sources of global ^{137}Cs contamination are nuclear weapon tests carried out in the Northern and Southern hemispheres in the 1950s and 1960s and the accident at the Chernobyl nuclear power plant in Ukraine on April 26, 1986. Thus, the highest ^{137}Cs peaks in sediments correspond to these periods (Abril and Garcia-Le6n, 1994; Erten, 1997; Di Leonardo et al., 2007; Erten, 2011) and serve as a validating tool in ^{210}Pb dating of sediments (Smith, 2001; Gunten et al., 2009).

2.6.3.5. *Challenges of quantifying sediment SARs using the ^{210}Pb dating models*

Contrary to the assumption of uniform sedimentation, non-uniform deposition of sediments may occur in reality and present a challenge to accurate dating of sediment cores using the common ^{210}Pb models. Episodic delivery of sediments and associated tracers into sedimentary systems through events such as flooding and turbidity currents may result in a significantly higher than normal rate of ^{210}Pb supply (Appleby, 1997), a situation that can place a restriction on ^{210}Pb dating of sediment cores (Appleby, 2008) since the underlying assumptions of constant concentration and flux for the CIC and CRS models are conflicted and lead to errors in the sediment chronology (Appleby, 1997; Marques Jr. et al., 2006). In situations where there have been significant variations in the sedimentation, both the CIC and CRS models are inapplicable for deriving chronologies (Appleby, 2008). In such situations, the CRS model may only be applicable if the sediment core is analysed in a piecewise fashion using ^{137}Cs and ^{241}Am chronostratigraphic dates as reference levels to derive accurate chronology for different sections of the core (Appleby, 2008).

Low activities of ^{137}Cs in sediments may compromise the reliability of the ^{210}Pb-derived sediment chronologies (Arnaud et al., 2006; Lepland et al., 2010). Moreover, in more complex depositional settings, the lack of a significant yearly deposition maximum of ^{137}Cs (Pfitzner et al., 2004) or the lack of distinct ^{137}Cs peaks (Abril and Garcia-Le6n, 1994) can result in a restricted use of ^{137}Cs in deriving sediment chronology and sedimentation rates. Therefore, the use of several independent sediment dating methods instead of only one method is recommended, since it can offset the limitations encountered when using only one dating method (Abril, 2004). In situations where different models (e.g. the CIC and CRS models) yield different results or neither of the results of the models is consistent with the ^{137}Cs results, the use of more robust models is recommended to aid the interpretation of the dating results (Smith, 2001; Appleby, 2004).

The assumption of no post-depositional mixing of sediments may also not be met in the real environment due to physical, chemical and biological (bioturbation) processes (Abril et al., 1992; Appleby, 1997; Schmidt et al., 2007b), which can potentially alter the original distribution of tracers in the sediments (Santschi et al., 2001) and cause difficulties in establishing accurate sediment chronology. The presence or absence of post-depositional sediment mixing can be assessed by examining the profiles of short-lived and particle-reactive radionuclides such as ^{234}Th (half-life = 24.1 d) and ^{7}Be (half-life = 53.3 d) in sediment cores (Erten, 1997; Schmidt et al., 2007a, b). ^{234}Th is produced from the decay of the soluble and long-lived parent ^{238}U (half-life = 4.5 x 10^9 y), while ^{7}Be is a cosmogenic radionuclide produced from the interaction of cosmic rays with oxygen and nitrogen in the stratosphere and the troposphere (Erten, 1997; Pfitzner et al., 2004; Mabit et al., 2014). ^{7}Be enters aquatic systems via direct wet and dry deposition or indirectly via transport of surface soil from the catchment through run-off (Pfitzner et al., 2004; Mabit et al., 2014). Due to their short half-lives, ^{234}Th and ^{7}Be can provide reliable information on sediment mixing occurring within a few months (Schmidt et al., 2007a). Subsurface occurrence of ^{234}Th and ^{7}Be in areas of low sedimentation rates may indicate post-depositional sediment mixing (Schmidt et al., 2007b).

The supply of ^{210}Pb$_{exc}$ and ^{137}Cs into bottom sediments is controlled by many complex processes (Abril et al., 1992; Abril and Garcia-Le6n, 1994; Appleby, 2008) that affect redistribution of sediment-bound tracers in confounding ways (Abril et al., 1992; Appleby, 2008; Mabit et al., 2014). For example, atmospheric ^{210}Pb and ^{137}Cs may first deposit onto soils and subsequently transported into aquatic systems (Abril and Garcia-Le6n, 1994; Mabit et al., 2014), in which case the flux of matter into the aquatic system will exert a greater control on the ^{210}Pb$_{exc}$ flux onto the bottom sediments than direct atmospheric ^{210}Pb$_{exc}$ deposition (Abril and Brunskill, 2014). Since the fluxes of matter is affected by storm runoff, floods and tidal events (Palinkas et al., 2005; Díaz-Asencio et al., 2009), which may be site- and time-specific, the application of a particular model at different sites would require a case by case assessment of sediment cores (Appleby and Oldfield, 1983; Abril and Brunskill, 2014). Moreover, the assumption of no post-depositional mobility of sediment-bound tracers under the application of ^{210}Pb dating models may also not hold in reality due to potential diffusion and advection of the tracers within the sediment pore water (Appleby and Garcia-Le6n, 1997) under the influence of the sediment porosity (Abril and Gharbi, 2012). An indication of post-depositional migration of sediment-bound radiotracers from surface layers into deeper layers through the sediment pores is the occurrence of maximum concentrations of radiotracers in subsurface layers rather than at the SWI (Abril and Gharbi, 2012).

2.7. Conclusions

Chemical pollution and sedimentation in the marine environment are major problems as they threaten human and ecological well-being and thus, their assessments have been the focus of many studies. Coastal harbours associated with industrial and urban centres are susceptible to chemical pollution and sedimentation. Chemical pollutants primarily bind to sediments upon entry into aquatic environments due to their particle-reactive nature, making sediments an important tool for chemical pollution assessment. Approaches such as radiological hazard indices and the ERICA model (for radionuclides), the enrichment factor, geo-accumulation index and the risk assessment code (for metals), biological effect-based numerical sediment quality guidelines (for e.g. metals, OCPs and PAHs) and the total toxicity equivalence can be used to provide a first line of evidence of chemical contamination in sediments. The inclusion of a battery of toxicity bioassays is now recommended as a further line of evidence to adequately characterise chemical pollution in sediments.

Quantifying sedimentation rates commonly involve the use of ^{210}Pb dating models such as the Constant Flux Constant Sedimentation, the Constant Initial Concentration and the Constant Rate of Supply models. In disturbed sedimentary systems such as coastal harbours where the delivery of sediments and ^{210}Pb may be controlled by complex processes, there is a potential limitation of the applicability of the ^{210}Pb dating models for the quantification of sedimentation rates. It is evident from the literature that integrated approaches are required to overcome such limitations to allow the quantification of sedimentation rates in non-ideal sedimentary environments. This requires more rigorous sampling and analytical schemes, the application of robust models, independent validation of the ^{210}Pb-derived chronology with other time markers such as ^{137}Cs and the use of multi-tracers with different half-lives, e.g. ^{7}Be and ^{234}Th, whose short half-lives can provide useful information on sedimentary processes over the very recent past (3-8 months).

References

Abrahim, G.M.S. & Parker, R.J. (2008). Assessment of heavy metal enrichment factors and the degree of contamination in marine sediments from Tamaki Estuary, Auckland, New Zealand. *Environ Monit Assess. 136*, 227-238

Abrajano Jr, T. A., Yan, B., Song, J., Bopp, R., & O'Malley, V. (2007). High-Molecular-Weight Petrogenic and Pyrogenic Hydrocarbons in Aquatic Environments. In D. H. Editors-in-Chief: Heinrich & K. T. Karl (Eds.), *Treatise on Geochemistry* (pp. 1-50). Oxford: Pergamon.

Abril, J.M. & Brunskill, G.J. (2014). Evidence that excess ^{210}Pb flux varies with sediment accumulation rate and implications for dating recent sediments. *J. Paleolimnol, 52*(3):121-137.

Abril J.M. & Gharbi, F. (2012). Radiometric dating of recent sediments: beyond the boundary conditions. J. Paleolimnol., *48*(2), 449-460.

Abril, J.M. (2004). Constraints on the use of ^{137}Cs as a time-marker to support CRS and SIT chronologies. *Environ. Pollut. 129*(1), 31-37.

Abril, J.M. & Garcia-Le6n, M. (1994). The Integrated Atmospheric Flux Effect in a Radiogeochronological Model. *J. Environ. Radioact. 24*, 65-79.

Abril, J.M., Garcia-Le6n, M., Garcia-Tenorio, R., Sinchez, C.I. & EI-Daoushy, F. (1992). Dating of Marine Sediments by an Incomplete Mixing Model. *J. Environ. Radioact. 15*, 135-151.

Adams, M.S. & Stauber, J.L. (2008). Marine whole sediment toxicity tests for use in temperate and tropical Australian environments: current status. *Aust. J. Ecotoxicol. 14*, 155-167.

Addo, M.A.; Darko, E.O.; Gordon, C.; Nyarko, B.J.B.; Gbadago, J.K.; Nyarko, E.; Affum, H.A.; Botwe, B.O. (2012). Evaluation of Heavy Metals Contamination of Soil and Vegetation in the Vicinity of a Cement Factory in the Volta Region, Ghana. *Int. J. Sci. and Technol., 2*(1): 40-50.

Addo, M.A.; Okley, G.M.; Affum, H.A.; Acquah, S.; Gbadago, J.K.; Senu, J.K.; Botwe, B.O. (2011). Water quality and level of some heavy metals in water and sediments of Kpeshie Lagoon, La-Accra, Ghana. *Res. J. Environ. Earth Sci., 3*(5): 487-497.

Ahmed, G., Anawar, H.M., Takuwa, D.T., Chibua, I.T., Singh, G.S. & Sichilongo, K. (2015). Environmental assessment of fate, transport and persistent behavior of dichlorodiphenyltrichloroethanes and hexachlorocyclohexanes in land and water ecosystems. *Int. J. Environ. Sci. Technol. 12*, 2741-2756.

Akrasi, S. (2011). Sediment discharges from Ghanaian rivers into the sea. *West Afr. J. Appl. Ecol. 18*, 1-13.

Alonso-Hernandez, C. M., Diaz-Asencio, M., Munoz-Caravaca, A., Delfanti, R., Papucci, C., Ferretti, O., & Crovato, C. (2006). Recent changes in sedimentation regime in Cienfuegos Bay, Cuba, as inferred from 210Pb and 137Cs vertical profiles. *Conti. Shelf Res., 26*(2), 153-167.

Aloupi, M. & Angelidis, M. (2001). Geochemistry of natural and anthropogenic metals in the coastal sediments of the island of Lesvos, Aegean Sea. *Environ. Pollut. 113*, 211-219.

Al-Trabulsy, H., Khater, A., & Habbani, F. (2011). Radioactivity levels and radiological hazard indices at the Saudi coastline of the Gulf of Aqaba. *Radiat. Phys. Chem. 80*(3), 343-348.

Alvarez-Vazquez, M.A., Caetano, M., Alvarez-Iglesias, P., del Canto Pedrosa-García, M., Calvo, S., De Una-Alvarez, E., Quintana, B., Vale, C. & Prego, R. (2017). Natural and anthropocene fluxes of trace elements in estuarine sediments of Galician Rias. *Estuar. Coast. Shelf Sci. 198*, 329-342.

Annicchiarico, C., Biandolino, F., Cardellicchio, N., Di Leo, A., Giandomenico, S., & Prato, E., 2007. Predicting toxicity in marine sediment in Taranto Gulf (Ionian Sea, Southern Italy) using Sediment Quality Guidelines and a battery bioassay. *Ecotoxicology, 16*, 239-246.

Appleby, P.G. (2008). Three decades of dating recent sediments by fallout radionuclides: A review. *Holocene, 18*(1), 83-93.

Appleby, P.G. & Oldfield, F. (1978). The calculation of lead-210 dates assuming a constant rate of supply of unsupported ^{210}Pb to the sediment. *CATENA, 5*(1), 1-8.

Appleby, P.G. (2004). Environmental change and atmospheric contamination on Svalbard: sediment chronology. *J. Paleolimnol., 31*(4), 433-444.

Appleby, P.G. & Oldfield, F. (1983). The assessment of ^{210}Pb data from sites with varying sediment accumulation rates. *Hydrobiologia, 103*(1), 29-35.

Appleby, P.G. (2002). Chronostratigraphic Techniques in Recent Sediments. In W. Last & J. Smol (Eds.), *Tracking Environmental Change Using Lake Sediments* (Vol. 1, pp. 171-203): Springer Netherlands.

Appleby, P.G. (1992). Dating recent sediments by [210]Pb: problems and solutions. STUK-A145.

Arnaud, F., Magand, O., Chapron, E., Bertrand, S., Boës, X., Charlet, F., & Mélières, M. A. (2006). Radionuclide dating (^{210}Pb, ^{137}Cs, ^{241}Am) of recent lake sediments in a highly active geodynamic setting (Lakes Puyehue and Icalma—Chilean Lake District). *Sci. Total Environ. 366*(2-3), 837-850.

Baran, A. & Tarnawski, M. (2013). Phytotoxkit/Phytotestkit and Microtox as tools for toxicity assessment of sediments. *Ecotoxicol. Environ. Safety 98*, 19-27.

Barneveld, H. & Hugtenburg, J. (2008). Feasibility study for implementation of sedimentation reduction measures in river harbours. RCEM, 1187-1192.

Bat, L. & Raffaelli, D. (1998). Sediment toxicity testing: a bioassay approach using the amphipod *Corophium volutator* and the polychaete *Arenicola marina*. *J. Exp. Mar. Bio. Ecol. 226*, 217-239.

Beg, K.R. & Ali, S. (2008). Microtox Toxicity Assay for the Sediment Quality Assessment of Ganga River. *American J. Environ. Sci. 4*(4), 383-387.

Beiras, R., Bellas, J., Fernández, N., Lorenzo, J. I. & Cobelo-García, A. (2003). Assessment of coastal marine pollution in Galicia (NW Iberian Peninsula); metal concentrations in seawater, sediments and mussels (*Mytilus galloprovincialis*) versus embryo–larval bioassays using *Paracentrotus lividus* and *Ciona intestinalis*. *Mar. Environ. Res. 56*(4), 531-553.

Bellucci, L. G., Giuliani, S., Romano, S., Albertazzi, S., Mugnai, C., & Frignani, M. (2012). An Integrated Approach to the Assessment of Pollutant Delivery Chronologies to Impacted Areas: Hg in the Augusta Bay (Italy). *Environ. Sci. Technol. 46*(4), 2040-2046.

Beresford, N., Brown, J., Copplestone, D., Garnier-Laplace, J., Howard, B., Larsson, C.-M., Oughton, D., Pröhl, G., & Zinger, I. (2007). D-ERICA: An integrated approach to the assessment and management of environmental risk from ionising radiation. Description of purpose, methodology and application.

Berto, D., Giani, M., Boscolo, R., Covelli, S., Giovanardi, O., Massironi, M., & Grassia, L. (2007). Organotins (TBT and DBT) in water, sediments, and gastropods of the southern Venice lagoon (Italy). *Mar. Pollut. Bull., 55*(10), 425-435.

Birch, G.F., & Hutson, P., 2009. Use of sediment risk and ecological/conservation value for strategic management of estuarine environments: Sydney estuary, Australia. *Environ. Manage. 44*, 836-850.

Birch, G.F., Gunns, T.J. & Olmos, M. (2015). Sediment-bound metals as indicators of anthropogenic change in estuarine environments. *Mar. Pollut. Bull. 101*, 243-257.

Birch, G.F. & Olmos, M.A. (2008). Sediment-bound heavy metals as indicators of human influence and biological risk in coastal water bodies. ICES *J. Mar. Sci*. *65*: 1407-1413.

Bloesch, J. & Burns, N. (1980). A critical review of sedimentation trap technique. *Schweiz Z Hydrol*., *42*(1), 15-55.

Botwe, B.O., Nyarko, E. & Ntow, W.J. (2012). Pesticide Contamination in Groundwater and Streams Draining Vegetable Plantations in the Ofinso District, Ghana. In: Hernandez-Soriano, M.C. (Ed.), Soil Health and Land Use Management, InTech, pp. 51-66. Rijeka, Croatia.

Brisson, L.N., Wolfe, D.A. & Staley, M. (2014). Interferometric swath bathymetry for large scale shallow water hydrographic surveys. Canadian Hydrographic Conference; 1-18.

Brucker, S., Clarke, J.H., Beaudoin, J., Lessels, C., Czotte, K., Loschiavo, R., Iwanowska, K. & Hill, P. (2007). Monitoring flood-related change in bathymetry and sediment distribution over the Squamish Delta, Howe Sound, British Columbia. U.S. Hydrographic Conference, 1-16.

Buesseler, K.O., Antia, A.N., Chen, M., Fowler, S.W., Gardner, W.D., Gustafsson, O., Harada, K., Michaels, A.F., Rutgers van der Loeff, M. & Sarin, M. (2007). An assessment of the use of sediment traps for estimating upper ocean particle fluxes. J Mar Res *65*(3), 345-416.

Burdige, D.J., 2007. Preservation of organic matter in marine sediments: controls, mechanisms, and an imbalance in sediment organic carbon budgets? *Chem. Rev. 107*(2), 467-485.

Burgess, R.M., Perron, M.M., Cantwell, M.G., Ho, K.T., Pelletier, M.C., Serbst, J.R. & Ryba, S.A. (2007). Marine sediment toxicity identification evaluation methods for the anionic metals arsenic and chromium. *Environ. Toxicol. Chem. 26*, 61-67.

Burton, G.A. (2002). Sediment quality criteria in use around the world. *Limnology*, *3*, 65-76.

Calmano, W., Hong, J. & Foerstner, U. (1993). Binding and mobilization of heavy metals in contaminated sediments affected by pH and redox potential. *Water. Sci. Technol*., *28*, 223-223.

Caroll, J. & Lerche, I. (2003). Sedimentary Processes: Quantification Using Radionuclides. Elsevier.

Carvalho, S., Cunha, M.R., Pereira, F., Pousão-Ferreira, P., Santos, M., & Gaspar, M. (2012). The effect of depth and sediment type on the spatial distribution of shallow soft-bottom amphipods along the southern Portuguese coast. *Helgol. Mar. Res. 66*, 489-501.

Casado-Martínez MC., Buceta JL, Belzunce MJ, DelValls TA (2006) Using sediment quality guidelines for dredged material management in commercial ports from Spain. *Environ. Int. 32*, 388-396.

Casado-Martinez, M.C., Forja, J.M., & DelValls, T.A., 2007. Direct comparison of amphipod sensitivities to dredged sediments from Spanish ports. *Chemosphere, 68*, 677-685.

Castro, Í.B., Arroyo, M.F., Costa, P.G. & Fillmann, G. (2012). Butyltin compounds and imposex levels in Ecuador. *Arch. Environ. Contam. Toxicol. 62*(1), 68-77.

Champ, P.M. & Seligman, P.F. (1996). Organotin: Environmental Fate and Effects, Chapman & Hall, London.

Chapman PM (2007) Determining when contamination is pollution - weight of evidence determinations for sediments and effluents. *Environ Int 33*(4), 492–501.

Chapman, P.M. & Anderson, J. (2005). A Decision-Making Framework for Sediment Contamination. *Integr. Environ. Assess. Manage., 1*(3): 163-173.

Chatterjee, M., Silva Filho E.V., Sarkar, S.K., Sella, S.M., Bhattacharya, A., Satpathy, K.K., Prasad, M.V.R., Chakraborty, S. & Bhattacharya, B.D. (2007) Distribution and possible source of trace elements in the sediment cores of a tropical macrotidal estuary and their ecotoxicological significance. *Environ. Int. 33*, 346-356.

Chen, C.Y., Wang, Y.J. and Yang, Ch. F. (2009). Estimating low toxic effect concentrations in closed system algal toxicity tests. *Ecotoxicol. Environ. Safety, 72*, 1514–1522.

Ciarelli, S., Vonck, W., Van Straalen, N., & Stronkhorst, J., 1998. Ecotoxicity assessment of contaminated dredged material with the marine amphipod *Corophium volutator. Arch. Environ. Contam. Toxicol. 34*, 350-356.

Clark, R.B. (2001). Marine Pollution. Oxford: Oxford University Press.

Clifton, R.J., Watson, P.G., Davey, J.T. & Frickers, P.E. (1995). A study of processes affecting the uptake of contaminants by intertidal sediments, using the radioactive tracers: [7]Be, ^{137}Cs and unsupported [210]Pb. *Estuar. Coast. Shelf Sci. 41*, 459-474.

Connon, R.E., Geist, J. & Werner, I. (2012). Effect-based tools for monitoring and predicting the ecotoxicological effects of chemicals in the aquatic environment. *Sensors*, **12**, 12741-12771.

Corbett, D.R., Walsh, J.P. & Marciniak, K. (2009). Temporal and spatial variability of trace metals in sediments of two adjacent tributaries of the Neuse River Estuary, North Carolina, USA. *Mar. Pollut. Bull., 58*(11), 1739-1747.

Davidson, C. M., Urquhart, G. J., Ajmone-Marsan, F., Biasioli, M., da Costa Duarte, A., Díaz-Barrientos, E., . . . Madrid, L. (2006). Fractionation of potentially toxic elements in urban soils from five European cities by means of a harmonised sequential extraction procedure. *Analytica Chimica Acta, 565*(1), 63-72.

de-la-Ossa-Carretero, J.A., Del-Pilar-Ruso, Y., Giménez-Casalduero, F., Sánchez-Lizaso, J.L. & Dauvin, J.-C. (2012). Sensitivity of amphipods to sewage pollution. *Estuar. Coastal Shelf Sci., 96*, 129-138

De Luca, G., Furesi, A., Leardi, R., Micera, G., Panzanelli, A., Piu, P.C. & Sanna, G. (2004). Polycyclic aromatic hydrocarbons assessment in the sediments of the Porto Torres Harbour (Northern Sardinia, Italy). *Mar. Chem. 86*(1), 15-32.

DelValls, T., Andres, A., Belzunce, M., Buceta, J., Casado-Martinez, M., Castro, R., Riba, I., Viguri, J.R. & Blasco, J. (2004). Chemical and ecotoxicological guidelines for managing disposal of dredged material. *Trends Anal. Chem. 23*, 819-828.

Denton, G. R. W., Concepcion, L. P., Wood, H. R. & Morrison, R. J. (2006). Polychlorinated biphenyls (PCBs) in sediments of four harbours in Guam. *Mar. Pollut. Bull. 52*(6), 711-718.

de Souza, V.L.B., Rodrigues, K.R.G., Pedroza, E. H., Melo, R.T., Lima, V.L., Hazin, C. A., . . Nascimento, R.K. (2012). Sedimentation Rate and ^{210}Pb Sediment Dating at Apipucos Reservoir, Recife, Brazil. *Sustainability, 4*(10), 2419-2429.

de Vicente, I., Cruz-Pizarro, L., Rueda, F.J. (2010). Sediment resuspension in two adjacent shallow coastal lakes: controlling factors and consequences on phosphate dynamics. *Aquat Sci, 72*(1):21-31.

de Vleeschouwer, F., Sikorski, J. & Fagel, N. (2010). Development of lead-210 measurement in peat using polonium extraction. A procedural comparison. *Geochronometria, 36*(1), 1-8.

Díaz-Asencio, M., Alonso-Hernández, C. M., Bolanos-Álvarez, Y., Gómez-Batista, M., Pinto, V., Morabito, R., . . . Sanchez-Cabeza, J. A. (2009). One century sedimentary record of Hg and Pb pollution in the Sagua estuary (Cuba) derived from ^{210}Pb and ^{137}Cs chronology. *Mar. Pollut. Bull., 59*(4–7), 108-115.

Di Leonardo, R., Bellanca, A., Capotondi, L., Cundy, A., & Neri, R. (2007). Possible impacts of Hg and PAH contamination on benthic foraminiferal assemblages: An example from the Sicilian coast, central Mediterranean. *Sci. Total Environ. 388*(1-3), 168-183.

Di Palma L, Mecozzi R (2007) Heavy metals mobilization from harbour sediments using EDTA and citric acid as chelating agents. *J. Hazard. Mater., 147*, 768-775.

Díaz-de Alba, M., Galindo-Riaño, M. D., Casanueva-Marenco, M. J., García-Vargas, M. & Kosore, C. M. (2011). Assessment of the metal pollution, potential toxicity and speciation of sediment from Algeciras Bay (South of Spain) using chemometric tools. *J. Hazard. Mater. 190*(1-3), 177-187.

Dung, T.T.T., Cappuyns, V., Swennen, R. & Phung, N.K. (2013). From geochemical background determination to pollution assessment of heavy metals in sediments and soils. *Rev Environ Sci Bio/Technol, 12*, 335-353.

Eggen, R.I., Behra, R., Burkhardt-Holm, P., Escher, B.I., & Schweigert, N., 2004. Challenges in ecotoxicology. *Environ. Sci. Technol. 38*, 58A-64A

El Mamoney, M., & Khater, A.E. (2004). Environmental characterization and radio-ecological impacts of non-nuclear industries on the Red Sea coast. *J. Environ. Radioact. 73*(2), 151-168.

Erten, H., Von Gunten, H., Rössler, E. & Sturm, M. (1985). Dating of sediments from Lake Zurich (Switzerland) with210Pb and ^{137}Cs. *Swiss J Hydrol, 47*(1), 5-11.

Erten H. (1997). Radiochronology of lake sediments. *Pure Appl Chem, 69*(1),71-76.

Erten, H.N. (2011). Radiochronological Methods as Tools to Study Environmental Pollution. In H. Gökçekus, U. Türker, & J.W. LaMoreaux (Eds.), *Survival and Sustainability* (pp. 1015-1022): Springer Berlin Heidelberg.

Escher, B.I., Bramaz, N., Mueller, J.F., Quayle, P., Rutishausera, S. & Vermeirssena, E.L.M (2008). Toxic equivalent concentrations (TEQs) for baseline toxicity and specific modes of action as a tool to improve interpretation of ecotoxicity testing of environmental samples. *J. Environ. Moni*t. *10*, 612-621.

Forrester, G.E., Fredericks, B.I., Gerdeman, D., Evans, B., Steele, M.A., Zayed, K., Schweitzer, L.E., Suffet, I.H., Vance, R.R., & Ambrose, R.F., 2003. Growth of estuarine fish is associated with the combined concentration of sediment contaminants and shows no adaptation or acclimation to past conditions. *Mar. Environ. Res. 56*, 423-442.

Gaillardet, J., Viers, J. & Dupre, B. (2003). Trace elements in river waters. In: Holland, H.D., Turekian, K.K. (Eds.), Treatise on Geochemistry, vol. 5. Elsevier, pp. 225-272.

Gaion, A., Sartori, D., Scuderi, A., & Fattorini, D. (2014). Bioaccumulation and biotransformation of arsenic compounds in *Hediste diversicolor* (Muller 1776) after exposure to spiked sediments. *Environ. Sci. Pollut. Res. 21*, 5952-5959.

Geissen, V., Mol, H., Klumpp, E., Umlauf, G., Nadal, M., van der Ploeg, M., van de Zee, S.E.A.T.M. & Ritsema, C.J. (2015). Emerging pollutants in the environment: A challenge for water resource management. *Int. Soil Water Conserv. Res. 3*, 57-65.

George, D., Gelfenbaum, G., & Stevens, A. (2012). Modeling the Hydrodynamic and Morphologic Response of an Estuary Restoration. *Estuaries and Coasts, 35*(6), 1510-1529.

Ghirardini, A.V., Girardini, M., Marchetto, D. & Pantani, C. (2009). Microtox solid phase test: Effect of diluent used in toxicity test. *Ecotoxicol. Environ. Safety, 72,* 851-861.

Giffin, D. & Corbett, D.R. (2003). Evaluation of sediment dynamics in coastal systems via short-lived radioisotopes. *J. Mar. Syst., 42*(3), 83-96.

Giuliani, S., Piazza, R., Bellucci, L. G., Cu, N. H., Vecchiato, M., Romano, S., . . . Frignani, M. (2011). PCBs in Central Vietnam coastal lagoons: Levels and trends in dynamic environments. *Mar. Pollut. Bull. 62*(5), 1013-1024.

Giuliani, S., Piazza, R., El Moumni, B., Polo, F.P., Vecchiato, M., Romano, S., Zambon, S., Frignani, M. & Bellucci, L.G. (2015). Recognizing different impacts of human and natural sources on the spatial distribution and temporal trends of PAHs and PCBs (including PCB-11) in sediments of the Nador Lagoon (Morocco). *Sci. Total Environ. 526,* 346-357.

Giuliani, S., Sprovieri, M., Frignani, M., Cu, N.H., Mugnai, C., Bellucci, L.G., Albertazzi, S., Romano, S., Feo, M.L., Marsella, E. and Nhon, D.H. (2008). Presence and origin of polycyclic aromatic hydrocarbon in sediments of nine coastal lagoons in central Vietnam. *Mar. Pollut. Bull. 56,* 1486-1512.

Giuliani, S., Piazza, R., Bellucci, L. G., Cu, N. H., Vecchiato, M., Romano, S., . . . Frignani, M. (2011). PCBs in Central Vietnam coastal lagoons: Levels and trends in dynamic environments. *Mar. Pollut. Bull., 62*(5), 1013-1024.

Gómez-Gutiérrez, A., Garnacho, E., Bayona, J. M. & Albaigés, J. (2007). Screening ecological risk assessment of persistent organic pollutants in Mediterranean sea sediments. *Environ. Int. 33*(7), 867-876.

Green, M.O. & Coco, G. (2014). Review of wave-driven sediment resuspension and transport in estuaries. *Rev. Geophys. 52*(1), 77-117.

Guerra, R., Dinelli, E., Righi, S., Forja, J.M. & García-Luque, E. (2010). Sediment geochemistry and accumulation rates on the northeastern shelf of the Gulf of Cádiz (SW Iberian Peninsula). *Adv. Mar. Chem*, DOI: 10.3989/scimar.2010.74s1067.

Guerra, R. (2012). Polycyclic Aromatic Hydrocarbons, Polychlorinated Biphenyls and Trace Metals in Sediments from a Coastal Lagoon (Northern Adriatic, Italy). *Water, Air, Soil Pollut. 223*(1*),* 85-98.

Gunten, L., Grosjean, M., Beer, J., Grob, P., Morales, A. & Urrutia, R. (2009). Age modeling of young non-varved lake sediments: methods and limits. Examples from two lakes in Central Chile. *J. Paleolimnol., 42*(3), 401-412.

Gupta, A.K., Gupta, S.K. & Rashmi, S.P. (2005). Environmental management plan for port and harbour projects. *Clean Techn Environ Policy. 7*, 133‑141.

Hakanson, L. (2006). Suspended Particulate Matter in lakes, rivers, and marine systems. Blackburn Press, Caldwell.

Hakanson, L., & Jansson, M. (1983) Principles of lake sedimentology. Springer Verlag, Berlin.

Hamed, M.A., Mohamedein, L. I., El-Sawy, M.A., & El-Moselhy, K.M. (2013). Mercury and tin contents in water and sediments along the Mediterranean shoreline of Egypt. *The Egypt. J. Aquat. Res., 39*(2), 75-81.

Hamzeh, M., Ouddane, B., Daye, M., Halwani, J. (2014). Trace Metal Mobilization from Surficial Sediments of the Seine River Estuary. *Water Air Soil Pollut. 225*, 1878-1892.

Hanna, S.K., Miller, R.J., Zhoud, D., Keller, A.A. & Lenihan, H.S. (2013). Accumulation and toxicity of metal oxide nanoparticles in a soft-sediment estuarine amphipod. *Aquatic Toxicology, 142-143*, 441- 446

Hannewijk, A., Kater, B.J. & Schipper, C.A. (2004). Sediment toxiciteitstest met *Nereis diversicolor* (In Dutch). 22 pp.

Hassona, R.K., Sam, A.K., Osman, O.I., Sirelkhatim, D.A. & LaRosa, J. (2008). Assessment of committed effective dose due to consumption of Red Sea coral reef fishes collected from the local market (Sudan). *Sci. Total Environ. 393*(2-3), 214-218.

Helali, M.A., Zaaboub, N., Oueslati, W., Added, A. & Aleya, L. (2016). Suspended particulate matter fluxes along with their associated metals, organic matter and carbonates in a coastal Mediterranean area affected by mining activities. *Mar. Pollut. Bull., 104*(1), 171-181.

Hernandez, J-M.A. (2015). Why would we use the Sediment Isotope Tomography (SIT) model to establish a ^{210}Pb-based chronology in recent-sediment cores? *J. Environ. Radioact., 143*, 40-46.

Hollins, S., Harrison, J., Jones, B., Zawadzki, A., Heijnis, H. & Hankin, S. (2011). Reconstructing recent sedimentation in two urbanised coastal lagoons (NSW, Australia) using radioisotopes and geochemistry. *J. Paleolimnol., 46*(4), 579-596.

Horowitz, A.J. & Elrick, K.A. (1987). The relation of stream sediment surface area, grain size and composition to trace element chemistry. *Applied Geochem. 2*(4), 437-451.

Horowitz, A. J., 1985. A primer on trace metal-sediment chemistry, U.S. Geological Survey Water-Supply Paper 2277, 67 p.

Hu, W., Wang, T., Khim, J.S., Luo, W., Jiao, W., Lu, Y., Naile, J.E., Chen, C., Zhang, X. & Giesy, J.P. (2009). HCH and DDT in Sediments from Marine and Adjacent Riverine Areas of North Bohai Sea, China. *Arch. Environ. Contam. Toxicol.* DOI 10.1007/s00244-009-9455-z.

ICES, 2008. Report of the Fourth ICES/OSPAR Workshop on Integrated Monitoring of Contaminants and their Effects in Coastal and Open Sea Areas (WKIMON IV). ICES Document CM 2008/ACOM: 49. 71 pp.

Iqbal, J., Tirmizi, S. & Shah, M. (2013). Statistical apportionment and risk assessment of selected metals in sediments from Rawal Lake (Pakistan). *Environ. Monit. Assess.* *185*(1), 729-743.

Islam, M.S. & Tanaka, M. (2004). Impacts of pollution on coastal and marine ecosystems including coastal and marine fisheries and approach for management: a review and synthesis. *Mar. Pollut. Bull.* *48*(7-8), 624-649.

Jain, C. (2004). Metal fractionation study on bed sediments of River Yamuna, India. *Water Res.*, 38, 569-578.

James, I. D. (2002). Modelling pollution dispersion, the ecosystem and water quality in coastal waters: a review. *Environ. Model. Software, 17*(4), 363-385.

James, I. D. (2002). Modelling pollution dispersion, the ecosystem and water quality in coastal waters: a review. *Environ. Model. Software, 17*(4), 363-385.

Jiang, J., Wang, J., Liu, S., Lin, C, He, M. & Liu, X. (2013). Background, baseline, normalization, and contamination of heavy metals in the Liao River Watershed sediments of China. *J. Asian Earth Sci. 73*, 87-94.

Kalman, J., Riba, I., DelValls, A. and Blasco, J. (2012). Bioaccumulation and effects of metals bound to sediments collected from Gulf of Cádiz (SW Spain) using the polychaete *Arenicola marina. Arch. Environ. Contam. Toxicol.*, *62*, 22–28.

Kalwa, M., Quináia, S., Pletsch, A., Techy, L., & Felsner, M. (2013). Fractionation and Potential Toxic Risk of Metals From Superficial Sediment in Itaipu Lake—Boundary Between Brazil and Paraguay. *Arch. Environ. Contam. Toxicol. 64*(1), 12-22.

Kannan, K., Johnson-Restrepo, B., Yohn, S. S., Giesy, J. P., & Long, D. T. (2005). Spatial and Temporal Distribution of Polycyclic Aromatic Hydrocarbons in Sediments from Michigan Inland Lakes. *Environ. Sci. Technol., 39*(13), 4700-4706.

Kelderman, P. (2012). Sediment Pollution, Transport, and Abatement Measures in the City Canals of Delft, The Netherlands. *Water Air Soil Pollut. 223*(7), 4627-4645.

Kelderman, P. & Osman, A.A. (2007). Effect of redox potential on heavy metal binding forms in polluted canal sediments in Delft (The Netherlands). *Water Res. 41*(18), 4251-4261.

Kelderman P, Ang'weya R, De Rozari P, Vijverberg T. (2012). Sediment characteristics and wind-induced sediment dynamics in shallow Lake Markermeer, the Netherlands. *Aquat Sci, 74*(2), 301-313.

Khaba, L. & Griffiths, J.A. (2017). Calculation of reservoir capacity loss due to sediment deposition in the Muela reservoir, Northern Lesotho. *Int Soil Water Conserv Res, 5,* 130-140.

King, A., Readman, J. & Zhou, J., 2004. Dynamic behaviour of polycyclic aromatic hydrocarbons in Brighton marina, UK. *Mar. Pollut. Bull. 48*(3), 229-239.

Koide, M., Bruland, K.W., & Goldberg, E.D. (1973). Th-228/Th-232 and Pb-210 geochronologies in marine and lake sediments. *Geochimica et Cosmochimica Acta, 37*(5), 1171-1187.

Kurnaz, A., Küçükömeroğlu, B., Keser, R., Okumusoglu, N., Korkmaz, F., Karahan, G., & Çevik, U. (2007). Determination of radioactivity levels and hazards of soil and sediment samples in Fırtına Valley (Rize, Turkey). *Appl. Radiat.Isot., 65*(11), 1281-1289.

Kusimi, J.M. & Dika, J.L. (2012). Sea erosion at Ada Foah: assessment of impacts and proposed mitigation measures. *Nat Hazards, 64,* 983-997.

Larsson, C.-M. (2008). An overview of the ERICA Integrated Approach to the assessment and management of environmental risks from ionising contaminants. *J. Environ. Radioact., 99*(9), 1364-1370.

Lepland, A., Andersen, T.J., Lepland, A., Arp, H.P.H., Alve, E., Breedveld, G.D., & Rindby, A., 2010. Sedimentation and chronology of heavy metal pollution in Oslo harbour, Norway. *Mar. Pollut. Bull. 60,* 1512-1522.

Leys, V., & Mulligan, R.P. (2011). Modelling coastal sediment transport for harbour planning: selected case studies. INTECH.

Libralato, G., Losso, C., Novelli, A.A., Citron, M., Sala, S.D., Zanotto, E., Cepak, F. & Ghirardini, A.V. (2008). Ecotoxicological evaluation of industrial port of Venice (Italy) sediment samples after a decontamination treatment. *Environ. Pollut. 156,* 644-650.

Lin, T., Hu, Z., Zhang, G., Li, X., Xu, W., Tang, J. & Li, J. (2009). Levels and mass burden of DDTs in sediments from fishing harbours: the importance of DDT-containing antifouling paint to the coastal environment of China. *Environ. Sci. Technol.* *43*(21), 8033-8038.

Little, M. (2003). Risks associated with ionizing radiation Environmental pollution and health. *Brit. Med. Bull.* *68*(1), 259-275.

Liu, J., Clift, P.D., Yan, W., Chen, Z., Chen, H., Xiang, R., & Wang, D. (2014). Modern transport and deposition of settling particles in the northern South China Sea: Sediment trap evidence adjacent to Xisha Trough. *Deep Sea Research Part I: Oceanographic Research Papers, 93*, 145-155.

Livingston, H.D. & Povinec, P.P. (2000). Anthropogenic marine radioactivity. *Ocean & Coastal Manage.* *43*(8), 689-712.

Long, E.R., Ingersoll, C.G. & MacDonald, D.D. (2006). Calculation and uses of mean sediment quality guideline quotients: a critical review. *Environ. Sci. Technol.* *40*(6), 1726-1736.

Long, E.R., Field, L.J., & MacDonald, D.D. (1998). Predicting toxicity in marine sediments with numerical sediment quality guidelines. *Environ. Toxicol. Chem., 17*(4), 714-727.

Long, E.R., MacDonald, D.D., Smith, S. L., & Calder, F.D. (1995). Incidence of adverse biological effects within ranges of chemical concentrations in marine and estuarine sediments. *Environ. Manage., 19*(1), 81-97.

Lu, X. (2007). A note on removal of the compaction effect for the ^{210}Pb method. *Appl. Radiat. Isot, 65*(1), 142-146.

Luo, X., Mai, B., Yang, Q., Fu, J., Sheng, G. & Wang, Z. (2004). Polycyclic aromatic hydrocarbons (PAHs) and organochlorine pesticides in water columns from the Pearl River and the Macao harbor in the Pearl River Delta in South China. *Mar. Pollut. Bull. 48*, 1102‑1115.

Luo, W., Lu, Y., Wang, T., Hu, W., Jiao, W., Naile, J.E., Khim, J.S. & Giesy, J.P. (2010). Ecological risk assessment of arsenic and metals in sediments of coastal areas of northern Bohai and Yellow Seas, China. *Ambio, 39*(5-6), 367-375.

Mabit, L., Benmansour, M. & Walling, D.E. (2008). Comparative advantages and limitations of the fallout radionuclides ^{137}Cs, ^{210}Pb$_{ex}$ and ^{7}Be for assessing soil erosion and sedimentation. *J. Environ. Radioact.* *99*, 1799-1807.

Mabit, L., Benmansour, M., Abril, J., Walling, D., Meusburger, K., Iurian, A.R., Bernard, C., Tarján, S., Owens, P.N., Blake, W.H., & Alewell, C. (2014). Fallout ^{210}Pb as a soil and sediment tracer in catchment sediment budget investigations: a review. *Earth-Sci. Rev., 138*, 335-351.

MacKenzie, A.B., Hardie, S.M.L., Farmer, J.G., Eades, L.J. & Pulford, I.D. (2011). Analytical and sampling constraints in ^{210}Pb dating. *Sci. Total Environ., 409*, 1298-1304.

Mahapatra, D.M., Chanakya H.N. & Ramachandra, T.V. (2011). C:N ratio of Sediments in a sewage fed Urban Lake. *Int. J. Geol., 5*(3), 86-92.

Mahu, E, Nyarko, E., Hulme, S. & Coale, K.H. (2015). Distribution and enrichment of trace metals in marine sediments from the Eastern Equatorial Atlantic, off the Coast of Ghana in the Gulf of Guinea. *Mar. Pollut. Bull., 98*, 301-307.

Magi, E. & Di Carro, M. (2016). Marine environment pollution: The contribution of mass spectrometry to the study of seawater. *Mass Spectrom. Rev.* doi: 10.1002/mas.21521.

Mai, B., Qi, S., Zeng, E.Y., Yang, Q., Zhang, G., Fu, J., Sheng, G., Peng, P. & Wang, Z. (2003). Distribution of polycyclic aromatic hydrocarbons in the coastal region off Macao, China: assessment of input sources and transport pathways using compositional analysis. *Environ. Sci. Technol., 37*, 4855-4863.

Mamindy-Pajany, Y., Bataillard, P., Séby, F., Crouzet, C., Moulin, A., Guezennec, A.-G., Hrel, C., Marmier, N., & Battaglia-Brunet, F. (2013). Arsenic in Marina Sediments from the Mediterranean Coast: Speciation in the Solid Phase and Occurrence of Thioarsenates. *Soil and Sediment Contam: An Int. J., 22*(8), 984-1002.

Mariño-Balsa, J.C., Pérez, P., Estévez-Blanco, P., Saco-Álvarez, L., Fernández, E. and Beiras, R. (2003). Assessment of the toxicity of sediment and seawater polluted by the Prestige fuel spill using bioassays with clams (Venerupis pollastra, Tappes decussates and Venerupis rhomboideus) and the microalga Skeletonema costatum. Ciencias Marinas *29*, 115–122.

Marques Jr, A. N., Monna, F., da Silva Filho, E. V., Fernex, F. E., & Fernando Lamego Simões Filho, F. (2006). Apparent discrepancy in contamination history of a subtropical estuary evaluated through 210Pb profile and chronostratigraphical markers. *Mar. Pollut. Bull., 52*(5), 532-539.

Marsden, I., & Rainbow, P., 2004. Does the accumulation of trace metals in crustaceans affect their ecology - the amphipod example? *J. Exp. Mar. Bio. Ecol. 300*, 373-408.

Mayor, D.J., Solan, M., Martinez, I., Murray, L., McMillan, H., Paton, G.I., & Killham, K., (2008). Acute toxicity of some treatments commonly used by the salmonid aquaculture industry to *Corophium volutator* and *Hediste diversicolor*: Whole sediment bioassay tests. *Aquaculture, 285*, 102-108.

McDonald, C. P., & Urban, N. R. (2007). Sediment radioisotope dating across a stratigraphic discontinuity in a mining-impacted lake. *J. Environ. Radioact., 92*(2), 80-95.

Mensink, B.P., Everaarts, J.M., Kralt, H., ten Hallers-Tjabbes, C.C. & Boon, J.P. (1996a). Tributyltin exposure in early life stages induces the development of male sexual characteristics in the common whelk, *Buccinum undatum*. *Mar. Environ. Res. 42*, 151-154.

Mensink, B.P., ten Hallers-Tjabbes, C.C., Kralt, J., Freriks, I.L. & Boon, J.B. (1996b). Assessment of imposex in the common whelk, *Buccinum undatum* (L.) from the Eastern Scheldt, the Netherlands. *Mar. Environ. Res. 41*, 315-325.

Mestres M, Sierra J, Mösso C. & Sánchez-Arcilla, A (2010). Sources of contamination and modelled pollutant trajectories in a Mediterranean harbour (Tarragona, Spain). *Mar. Pollut. Bull. 60*, 898-907.

Meybeck, M., Vörösmarty, C., 2005. Fluvial filtering of land-to-ocean fluxes: from natural Holocene variations to Anthropocene. *Comptes Rendus Geosci. 337*, 107-123.

Morales-Caselles, C., Riba, I. & DelValls, T.A (2009). A weight of evidence approach for quality assessment of sediments impacted by an oil spill: The role of a set of biomarkers as a line of evidence. *Mar. Environ. Res. 67*, 31-37.

Morales-Caselles, C., Riba, I., Sarasquete, C. & DelValls, T.A. (2008). The application of a weight of evidence approach to compare the quality of coastal sediments affected by acute (Prestige 2002) and chronic (Bay of Algeciras) oil spills. *Environ. Pollut. 156*, 394-402.

Morales-Caselles, C., Kalman, J., Riba, I. & DelValls, T.A. (2007). Comparing sediment quality in Spanish littoral areas affected by acute (Prestige, 2002) and chronic (Bay of Algeciras) oil spills. *Environ. Pollut., 146*, 233-240.

Moreira, S.M., Lima, I., Ribeiro, R., & Guilhermino, L. (2006). Effects of estuarine sediment contamination on feeding and on key physiological functions of the polychaete *Hediste diversicolor*: laboratory and in situ assays. *Aquatic Toxicol. 78*, 186-201.

Moukhchan, F., March, J. G., & Cerdá, V. (2013). Distribution of trace metals in marine sediments of the Bay of Palma de Mallorca (Mallorca Island, Spain). *Environ. Monit. Assess. 185*(1), 695-706.

Muller, G., 1969. Index of geo-accumulation in sediments of the Rhine River. *J. Geol., 2*(3): 108-118.

Mulligan, C.N., Yong, R.N. & Gibbs, B.F. (2001). An evaluation of technologies for the heavy metal remediation of dredged sediments. *J. Hazard. Mater. 85*, 145-163.

Neff, J.M., Stout, S.A. & Gunster, D.G. (2005). Ecological risk assessment of polycyclic aromatic hydrocarbons in sediments: identifying sources and ecological hazard. *Integr. Environ. Assess. Manage. 1*(1), 22-33.

Nehyba, S., Nývlt, D., Schkade, U., Kirchner, G. & Franců, E. (2011). Depositional rates and dating techniques of modern deposits in the Brno reservoir (Czech Republic) during the last 70 years. *J. Paleolimnol., 45*(1), 41-55.

Nisbet, I.C. & LaGoy, P.K., 1992. Toxic equivalency factors (TEFs) for polycyclic aromatic hydrocarbons (PAHs). *Regul. Toxicol. Pharm. 16*(3), 290-300.

Ntow, W.J. & Botwe, B.O. (2011). Contamination status of organochlorine pesticides in Ghana. In: Loganathan, B.G. & Lam, P.K.S. (Eds), Global Contamination Trends of Persistent Organic Chemicals. pp. 393-411. CRC Press, USA.

Nyarko, E., Botwe, B.O., Lamptey, E., Nuotuo, O., Foli, B.A., Addo, M.A. (2011a). Toxic metal concentrations in deep-sea sediments from the jubilee oil field and surrounding areas off the western coast of Ghana. *Trop. Environ. Res. 9*, 584-595.

Nyarko, E., Botwe, B., Ansong, J., Delfanti, R., Barsanti, M., Schirone, A. & Delbono, I. (2011b). Determination of ^{210}Pb, ^{226}Ra and ^{137}Cs in beach sands along the coastline of Ghana. *Afr. J. Environ. Pollut. Health, 9*, 17-23.

Nyarko, E., Botwe, B.O. & Klubi, E. (2011c). Polycyclic aromatic hydrocarbons (PAHs) levels in two commercially important fish species from coastal waters of Ghana and their carcinogenic health risks. *West Afr. J. Appl. Ecol., 19*, 53-66.

Nyarko, E., Fletcher, A., Addo, S., Foli, B.A.K., Mahu, E. (2014). Geochemical Assessment of Heavy Metals in Surface Sediments: A Case Study of the Tema Port, Ghana. *J. Ship. Ocean Eng. 4*, 79-92.

Oehlmann, J. (2002). Baseline study on intersex in Littorina littorea with recommendations for biological TBT assessment criteria. Report, Johann Wolfgang Goethe University, Frankfurt am Main. SIME 02/6/1-E.

Oehlmann, J. (2002). Baseline study on intersex in Littorina littorea with recommendations for biological TBT assessment criteria. Report, Johann Wolfgang Goethe University, Frankfurt am Main. SIME 02/6/1-E.

Oldfield, F. & Appleby, P.G. (1984). A combined radiometric and mineral magnetic approach to recent geochronology in lakes affected by catchment disturbance and sediment redistribution. *Chem. Geol., 44*(1-3), 67-83.

Ortt, R.A., Kerhin, R.T., Wells, D. & Cornwell, J. (2000). Bathymetric survey and sedimentation analysis of Loch Raven and Prettyboy reservoirs. Coastal and Estuarine Geology File Report; 99-4.

Pagano, G., Guida, M., Trifuoggi, M., Thomas, P., Palumbo, A., Romano, G. and Oral, R. (2017a). Sea Urchin Bioassays in Toxicity Testing: I. Inorganics, Organics, Complex Mixtures and Natural Products. *Expert Opin Environ Biol.*, 6:1. doi: 10.4172/2325-9655.1000142

Pagano, G., Thomas, P., Guida, M., Palumbo, A., Romano, G., Oral, R. Trifuoggi, M. (2017b). Sea Urchin Bioassays in Toxicity Testing: II. *Sediment Eval. Expert Opin. Environ Biol.*, *6*, 1.

Palinkas, C., Nittrouer, C., Wheatcroft, R., Langone, L. (2005). The use of [7]Be to identify event and seasonal sedimentation near the Po River delta, Adriatic Sea. *Mar. Geol. 222*, 95-112.

Petrosillo, I., Valente, D., Zaccarelli, N. & Zurlini, G. (2009). Managing tourist harbors: Are managers aware of the real environmental risks? *Mar. Pollut. Bull.* 58, 1454-1461.

Pfitzner, J., Brunskill, G. & Zagorskis, I. (2004). [137]Cs and excess ^{210}Pb deposition patterns in estuarine and marine sediment in the central region of the Great Barrier Reef Lagoon, north-eastern Australia. *J. Environ. Radioact. 76*(1), 81-102.

Philippe, S., Leterme, C., Lesourd, S., Courcot, L., Haack, U. & Caillaud, J. (2008). Bioavailability of sediment-borne lead for ragworms (*Hediste diversicolor*) investigated by lead isotopes. *Appl. Geochem. 23*, 2932-2944.

Pini J, Richir J, Watson G (2015) Metal bioavailability and bioaccumulation in the polychaete Nereis (Alitta) virens (Sars): The effects of site-specific sediment characteristics. *Mar. Pollut. Bull.*, *95*, 565-575.

Power, E.A. & Chapman, P.M. (1992). Assessing sediment quality. In: G.A. Burton Jr. (Ed.), Sediment toxicity assessment. Lewis Publishers Inc, Boca raton, FL, 457pp.

Prato, E., Parlapiano, I. & Biandolino, F. (2011). Evaluation of a bioassays battery for ecotoxicological screening of marine sediments from Ionian Sea (Mediterranea Sea, Southern Italy). *Environ Monit Assess*, DOI 10.1007/s10661-011-2335-9.

Prato, E., Parlapiano, I. & Biandolino, F. (2015): Ecotoxicological evaluation of sediments by battery bioassays: application and comparison of two integrated classification systems, Chemistry and Ecology, DOI:10.1080/02757540.2015.1069278.

Ptistišek, N., Milačič, R., & Veber, M. (2001). Use of the BCR three-step sequential extraction procedure for the study of the partitioning of Cd, Pb and Zn in various soil samples. *J. Soils Sediments, 1*(1), 25-29.

Pueyo M, Rauret G, Lück D, Yli-Halla M, Muntau H, Quevauviller P, Lopez-Sanchez J (2001) Certification of the extractable contents of Cd, Cr, Cu, Ni, Pb and Zn in a freshwater sediment following a collaboratively tested and optimised three-step sequential extraction procedure. *J. Environ. Monit., 3*, 243-250.

Qu, W. & Kelderman, P. (2001) Heavy metal contents in the Delft canal sediments and suspended solids of the River Rhine: multivariate analysis for source tracing. *Chemosphere, 45*, 919-925.

Quensen III, J.F., Mueller, S.A., Jain, M.K. & Tiedje, J.M. (1998). Reductive Dechlorination of DDE to DDMU in Marine Sediment Microcosms. *Science, 280* (5364), 722-724.

Quevauviller, P., Rauret, G., López-Sánchez, J. F., Rubio, R., Ure, A., & Muntau, H. (1997). Certification of trace metal extractable contents in a sediment reference material (CRM 601) following a three-step sequential extraction procedure. *Sci. Total Environ., 205*(2–3), 223-234.

Quevauviller, P., Rauret, G., Muntau, H., Ure, A. M., Rubio, R., López-Sánchez, J. F., . . . Griepink, B. (1994). Evaluation of a sequential extraction procedure for the determination of extractable trace metal contents in sediments. *Fresenius' J. Anal. Chem., 349*(12), 808-814.

Rae, J.E. (1997). Trace metals in deposited intertidal sediments. In: Jickells, T.D., Rae, J.E. (Eds.), Biogeochemistry of Intertidal Sediments, Cambridge Environmental Chemistry Series No. 9. Cambridge University Press, pp. 16-41.

Rainbow, P.S. (2007). Trace metal bioaccumulation: Models, metabolic availability and toxicity. *Environ. Int., 33*, 576-582.

Rainbow, P.S. & Luoma, S.N. (2011). Metal toxicity, uptake and bioaccumulation in aquatic invertebrates – Modelling zinc in crustaceans. *Aquatic Toxicol. 105,* 455-465

Rajendran, R. B., Imagawa, T., Tao, H., & Ramesh, R. (2005). Distribution of PCBs, HCHs and DDTs, and their ecotoxicological implications in Bay of Bengal, India. *Environ. Int., 31*(4), 503-512.

Rani, M., Shanker, U. & Jassal, V. (2017). Recent strategies for removal and degradation of persistent & toxic organochlorine pesticides using nanoparticles: A review. *J. Environ. Manage. 190*, 208-222.

Ravanat, J., Breton, J., Douki, T., Gasparutto, D., Grand, A., Rachidi, W., & Sauvaigo, S. (2014). Radiation-mediated formation of complex damage to DNA: a chemical aspect overview. *Brit. J. Radiol.* *87*(1035), 20130715.

Rauret, G., Lopez-Sanchez, J.F., Sahuquillo, A., Rubio, R., Davidson, C., Ure, A. & Quevauviller, P. (1999). Improvement of the BCR three step sequential extraction procedure prior to the certification of new sediment and soil reference materials. *J. Environ. Monit., 1*, 57-61.

Ré, A., Freitas, R., Sampaio, L., Rodrigues, A.M., & Quintino, V. (2009). Estuarine sediment acute toxicity testing with the European amphipod *Corophium multisetosum* Stock, 1952. *Chemosphere, 76*, 1323-1333.

Robbins, J.A. (1978). Geochemical and geophysical applications of radioactive lead isotopes. In: Nriagu JO (Ed.), Biogeochemistry of Lead. Elsevier.

Rocher, V., Azimi, S., Moilleron, R., & Chebbo, G. (2004). Hydrocarbons and heavy metals in the different sewer deposits in the 'Le Marais' catchment (Paris, France): stocks, distributions and origins. *Sci. Total Environ., 323*(1-3), 107-122.

Roddie, B., & Thain, J. (2002). Biological effects of contaminants: *Corophium sp.* sediment bioassay and toxicity test: International Council for the Exploration of the Sea. ICES Techniques in Marine Environmental Sciences No. 28, ISSN 0903-2606, Copenhagen, .Denmark.

Rose, A., Carruthers, A-M, Stauber, J., Lima, R & Blockwell, S. (2006). Development of an acute toxicity test with the marine copepod Acartia Sinjiensis. *Australasian J. Ecotoxicol, 12*, 73-81.

Ruiz-Fernández, A.C., Frignani, M., Hillaire-Marcel, C., Ghaleb, B., Arvizu, M.D., Raygoza-Viera, J.R. & Páez-Osuna, F. (2009). Trace metals (Cd, Cu, Hg, and Pb) accumulation recorded in the intertidal mudflat sediments of three coastal lagoons in the Gulf of California, Mexico. *Estuar. Coasts.* DOI 10.1007/s12237-009-9150-3.

Ryding, S.-O. (1985). Chemical and Microbiological Processes as Regulators of the Exchange of Substances between Sediments and Water in Shallow Eutrophic Lakes. *Internationale Revue der gesamten Hydrobiologie und Hydrographie, 70*(5), 657-702.

Santos-Echeandía, J., Prego, R., Cobelo-García, A., & Caetano, M. (2012). Metal composition and fluxes of sinking particles and post-depositional transformation in a ria coastal system (NW Iberian Peninsula). *Mar. Chem., 134*, 36-46.

Santschi, P.H., Murray, J.W., Baskaran, M., Benitez-Nelson, C.R., Guo, L.D., Hung, C.C., Lamborg, C., Moran, S.B., Passow, U. & Roy-Barman, M. (2006). Thorium speciation in seawater. *Mar. Chem. 100*(3-4), 250-268.

Santschi, P.H., Presley, B.J., Wade, T.L., Garcia-Romero, B., & Baskaran, M. (2001). Historical contamination of PAHs, PCBs, DDTs, and heavy metals in Mississippi River Delta, Galveston Bay and Tampa Bay sediment cores. *Mar. Environ. Res. 52*(1), 51-79.

Sauvé, S. & Desrosiers, M. (2014). A review of what is an emerging contaminant. *Chem Centr J. 8*, 15.

Scaps, P. (2002). A review of the biology, ecology and potential use of the common ragworm *Hediste diversicolor* (OF Müller)(Annelida: Polychaeta). *Hydrobiologia, 470*, 203-218.

Scarlett, A., Rowland, S., Canty, M., Smith, E., & Galloway, T. (2007). Method for assessing the chronic toxicity of marine and estuarine sediment-associated contaminants using the amphipod *Corophium volutator. Mar. Environ. Res. 63*, 457-470.

Schettler, G., Mingram, J., Negendank, J.F.W., Jiaqi, L. (2006a). Paleovariations in the East-Asian Monsson regime geochemically recorded in varved sediments of Lake Sihailongwan (Northeast China, Jilin province). Part 2: a 200-year record of atmospheric lead-210 flux variations and its palaeoclimatic implications. *J. Paeolimnol. 35*, 271-288.

Schettler, G., Qiang, L., Mingram, J., Negendank, J.F.W. (2006b). Paleovariations in the East-Asian Monsson regime geochemically recorded in varved sediments of Lake Sihailongwan (Northeast China, Jilin province). Part 1:Hydrological conditions and flux. *J. Paleolimnol. 35*, 239-270.

Schipper, C., Rietjens, I., Burgess, R., Murk, A. (2010). Application of bioassays in toxicological hazard, risk and impact assessments of dredged sediments. *Mar. Pollut. Bull. 60*(11), 2026-2042.

Schipper, C.A., Smit, M.G.D., Kaag, N.H.B.M. & Vethaak, A.D. (2008). A weight-of-evidence approach to assessing the ecological impact of organotin pollution in Dutch marine and brackish waters; combining risk prognosis and fi eld monitoring using common periwinkles (*Littorina littorea*). *Mar. Environ. Res. 66*, 23-239.

Schipper, C.A., Burgess, R.M., van den Dikkenberg, L.C. (2006). Sediment toxiciteittest met de slijkgarnaal *Corophium volutator* (In Dutch). 18 pp.

Schmid, E., & Schrader, T. (2007). Different biological effectiveness of ionising and non-ionising radiations in mammalian cells. *Adv. Radio Sci., 5*(1), 1-4.

Schmidt S, Jouanneau J-M, Weber O, Lecroart P, Radakovitch O, Gilbert F, Jézéquel D. (2007a). Sedimentary processes in the Thau Lagoon (France): from seasonal to century time scales. *Estuar Coast Shelf Sci., 72*(3), 534-542.

Schmidt S, Gonzalez J-L, Lecroart P, Tronczyński J, Billy I, Jouanneau J-M. (2007b). Bioturbation at the water-sediment interface of the Thau Lagoon: impact of shellfish farming. *Aquat Living Resour., 20*(2), 163-169.

Sikorski, J. & Bluszcz, A. (2008). Application of α and γ spectrometry in the ^{210}Pb method to model sedimentation in artificial retention reservoir. *Geochronometria, 31*, pp. 65).

Simpson, C.D., Mosi, A.A., Cullen, W.R. & Reimer, K.J., 1996. Composition and distribution of polycyclic aromatic hydrocarbon contamination in surficial marine sediments from Kitimat Harbour, Canada. *Sci. Total Environ. 181*(3), 265-278.

Simboura, N. & Zenetos, A. (2002). Benthic indicators to use in Ecological Quality classification of Mediterranean soft bottom marine ecosystems, including a new Biotic Index. *Mediter. Mar. Sci. 3*(2), 77-111.

Sinex, S.A. & Helz, G.R. (1981). Regional Geochemistry of Trace Elements in Chesapeake Bay Sediments. *Environ. Geol.* **3**, 315-323.

Singh, A., & Lal, R. (2009). *Sphingobium ummariense* sp. nov., a hexachlorocyclohexane (HCH)-degrading bacterium, isolated from HCH-contaminated soil. *Int. J. Syst. Evol. Microbiol. 59*(1), 162-166.

Sirelkhatim, D., Sam, A., & Hassona, R. (2008). Distribution of ^{226}Ra–^{210}Pb–^{210}Po in marine biota and surface sediments of the Red Sea, Sudan. *J. Environ. Radioact. 99*(12), 1825-1828.

Smith, J., Lee, K., Gobeil, C. & Macdonald, R. (2009). Natural rates of sediment containment of PAH, PCB and metal inventories in Sydney Harbour, Nova Scotia. *Sci. Total Environ. 407*(17), 4858-4869.

Smith, J. (2001). Why should we believe ^{210}Pb sediment geochronologies? *J. Environ. Radioact., 55*(2), 121-123.

Souza, A. J. & Lane, A. (2013). Effects of freshwater inflow on sediment transport. *J. Operational Oceanography, 6*(1), 27-31.

Srivastava, A. and Shivanandappa, T. (2010). Stereospecificity in the cytotoxic action of hexachlorocyclohexane isomers. *Chemico-Biological Interactions, 183* (1): 34-39.

Stoschek, O. & Zimmermann, C. (2006). Water Exchange and Sedimentation in an Estuarine Tidal Harbor Using Three-Dimensional Simulation. *J. Waterway Port Coast. Ocean Engin. 132*(5), 410-414.

Stronkhorst, J., Schipper, C., Brils, J., Dubbeldam, M., Postma, J., & van de Hoeven, N. (2003). Using marine bioassays to classify the toxicity of Dutch harbor sediments. *Environ. Toxicol. Chem. 22*, 1535-1547.

Stronkhorst, J., van Hattum, B., Bowmers, T. (1999). Bioaccumulation and toxicity of tributyltin to a burrowing heart urchin and an amphipod in spiked, silt marine sediments. *Environ. Toxicol. Chem., 18*, 2343-2351.

Sugandhi, S., Joshi, V. M. & Ravi, P. (2014). Studies on natural and anthropogenic radionuclides in sediment and biota of Mumbai Harbour Bay. *J. Radioanal. Nucl. Chem. 300*(1), 67-70.

Sundberg, H., Tjärnlund, U., Åkerman, G., Blomberg, M., Ishaq, R., Grunder, K., . . . Balk, L. (2005). The distribution and relative toxic potential of organic chemicals in a PCB contaminated bay. *Mar. Pollut. Bull. 50*(2), 195-207.

Syvitski, J.P., Vörösmarty, C.J., Kettner, A.J. & Green, P. (2005). Impact of humans on the flux of terrestrial sediment to the global coastal ocean. *Science, 308*(5720), 376-380.

Szmytkiewicz, A., & Zalewska, T. (2014). Sediment deposition and accumulation rates determined by sediment trap and ^{210}Pb isotope methods in the Outer Puck Bay (Baltic Sea). *Oceanologia, 56*(1), 85-106.

Tang, C. W.-y., Ip, C. C.-m., Zhang, G., Shin, P. K. S., Qian, P.-y., & Li, X.-d. (2008). The spatial and temporal distribution of heavy metals in sediments of Victoria Harbour, Hong Kong. *Mar. Pollut. Bull., 57*(6-12), 816-825.

Tanner, P.A., Leong, L.S, & Pan, S.M. (2000). Contamination of heavy metals in marine sediment cores from Victoria Harbour, Hong Kong. *Mar. Pollut. Bull. 40*(9), 769-779.

Ten Hallers-Tjabbes, C.C., Kemp, J.F. & Boon, J.P. (1994). Imposex in whelks (*Buccinum undatum*) from the open North Sea: Relation to shipping traffic intensities. *Mar. Pollut. Bull. 28*, 311-313.

Tesi, T., Miserocchi, S., Goñi, M.A., Langone, L., Boldrin, A. & Turchetto, M. (2007). Organic matter origin and distribution in suspended particulate materials and surficial sediments from the western Adriatic Sea (Italy). *Estuar. Coast. Shelf Sci. 73*(3), 431-446.

Tessier, A., Campbell, P. G. C., & Bisson, M. (1979). Sequential extraction procedure for the speciation of particulate trace metals. *Anal. Chem., 51*(7), 844-851.

Thain, J., & Bifield, S. (2002). Biological effects of contaminants: sediment bioassay using the polychaete *Arenicola marina*: International Council for the Exploration of the Sea.

Thomas, M., Lazartigues, A., Banas, D., Brun-Bellut, J. & Feidt, C. (2102). Organochlorine pesticides and polychlorinated biphenyls in sediments and fish from freshwater cultured fish ponds in different agricultural contexts in north-eastern France. *Ecotoxicol. Environ. Safety*, *77*, 35-44.

Ulanovsky, A., Pröhl, G., & Gómez-Ros, J. (2008). Methods for calculating dose conversion coefficients for terrestrial and aquatic biota. *J. Environ. Radioact.*, *99*(9), 1440-1448.

UNSCEAR. (2000). *Sources and effects of ionizing radiation: sources (Vol. 1). New York*: United Nations Publications.

Uwadiae, R.E. (2010). An inventory of the benthic macrofauna of Epe Lagoon, southwest Nigeria. *J. Sci. Res. Develop. 12*, 161-171.

van Dam, R.A., Harford, A.J., Houston, M.A., Hogan, A.C. & Negri, A.P. (2008). Tropical marine toxicity testing in australia: a review and recommendations. *Australasian J. Ecotoxicol.*, *14*, 55-88.

van den Heuvel-Greve, M., Postma, J., Jol, J., Kooman, H., Dubbeldam, M., Schipper, C., & Kater, B. (2007). A chronic bioassay with the estuarine amphipod *Corophium volutator*: Test method description and confounding factors. *Chemosphere*, *66*, 1301-1309.

van Hullebusch, E.D., Lens, P.N.L. & Tabak, H.H. (2005). Developments in bioremediation of soils and sediments polluted with metals and radionuclides. 3. Influence of chemical speciation and bioavailability on contaminants immobilization/mobilization bio-processes. *Rev. Environ. Sci. Biotechnol.*, *4*(3), 185–212.

Van Rijn, L.C. (2005). Estuarine and coastal sedimentation problems. *Int. J. Sediment Res. 20*(1):39-51.

Vikas, M., & Dwarakish, G.S. (2015). Coastal Pollution: A Review. *Aquat. Procedia*, *4*, 381-388.

Volesky, B. (2001). Detoxification of metal-bearing effluents: biosorption for the next century. *Hydrometallurgy. 59*, 203-216.

Walker, K., Vallero, D.A. & Lewis, R.G., 1999. Factors influencing the distribution of lindane and other hexachlorocyclohexanes in the environment. *Environ. Sci. Technol. 33*(24), 4373-4378.

Wang, S., Jia, Y., Wang, S., Wang, X., Wang, He., Zhao, Z. & Liu, B. (2010). Fractionation of heavy metals in shallow marine sediments from Jinzhou Bay, China. *J. Environ. Sci. 22*(1), 23-31.

Wang, Z., Liu, Z., Xu, K., Mayer, L.M., Zhang, Z., Kolker, A.S. and Wu, W. (2014). Concentrations and sources of polycyclic aromatic hydrocarbons in surface coastal sediments of the northern Gulf of Mexico. *Geochem. Transact. 15*, 2.

Wang, X.-C., Zhang, Y.-X. & Chen, R.F. (2001). Distribution and partitioning of polycyclic aromatic hydrocarbons (PAHs) in different size fractions in sediments from Boston Harbor, United States. *Mar. Pollut. Bull. 42*(11), 1139-1149.

WHO (World Health Organisation) (1989). DDT and its derivatives: Environmental aspects. Environmental Health Criteria 83. Geneva.

Willett, K.L., Ulrich, E.M. & Hites, R.A. (1998). Differential toxicity and environmental fates of hexachlorocyclohexane isomers. *Environ. Sci. Technol.* **32**(15), 2197-2207.

Williams, C. (1996). Combating marine pollution from land-based activities: Australian initiatives. *Ocean & Coastal Manage. 33*(1-3), 87‑112.

Xinwei, L., Lingqing, W., & Xiaodan, J. (2006). Radiometric analysis of Chinese commercial granites. *J. Radioanal Nucl. Chem., 267*(3), 669-673.

Yan, B., Abrajano, T. A., Bopp, R. F., Benedict, L. A., Chaky, D. A., Perry, E., . . . Keane, D. P. (2006). Combined application of δ13C and molecular ratios in sediment cores for PAH source apportionment in the New York/New Jersey harbor complex. *Org. Geochem., 37*(6), 674-687.

Yang, R.-q., Jiang, G.-b., Zhou, Q.-f., Yuan, C.-g. & Shi, J.-b. (2005). Occurrence and distribution of organochlorine pesticides (HCH and DDT) in sediments collected from East China Sea. *Environ. Int., 31*(6), 799-804.

Yang, Y., Woodward, L.A., Li, Q.X. & Wang, J. (2014). Concentrations, Source and Risk Assessment of Polycyclic Aromatic Hydrocarbons in Soils from Midway Atoll, North Pacific Ocean. *PLoS One, 9*(1): 1-7.

Yeager, K., Santschi, P., Phillips, J. & Herbert, B. (2005). Suspended sediment sources and tributary effects in the lower reaches of a coastal plain stream as indicated by radionuclides, Loco Bayou, Texas. *Environ. Geol. 47*(3), 382-395.

Yilgor, S., Kucuksezgin, F., & Ozel, E. (2012). Assessment of metal concentrations in sediments from Lake Bafa (Western Anatolia): an index analysis approach. *Bulletin of Environ. Contam. Toxicol., 89*(3), 512-518.

Yim, U.H., Hong, S.H. & Shim, W.J. (2007). Distribution and characteristics of PAHs in sediments from the marine environment of Korea. *Chemosphere*, *68*, 85-92

Yunker, M.B., Macdonald, R.W., Brewer, R., Sylvestre, S., Tuominen, T., Sekela, M., Mitchell, R.H., Paton, D.W., Fowler, B.R., Gray, C., Goyette, D. & Sullivan, D. (2000). Assessment of natural and anthropogenic hydrocarbon inputs using PAHs as tracers. The Fraser River basin and Strait of Georgia, 1987-1997. Burrard Inlet Environmental Action Programme, Canada.

Yunker, M.B., Macdonald, R.W., Vingarzan, R., Mitchell, R.H., Goyette, D. & Sylvestre, S. (2002). PAHs in the Fraser River basin: a critical appraisal of PAH ratios as indicators of PAH source and composition. *Org. Geochem. 33*(4), 489-515.

Zaborska, A., Carroll, J., Papucci, C., & Pempkowiak, J. (2007). Intercomparison of alpha and gamma spectrometry techniques used in ^{210}Pb geochronology. *J. Environ. Radioact., 93*(1), 38-50.

Zhou, J., Ma, D., Pan, J., Nie, W., & Wu, K. (2008). Application of multivariate statistical approach to identify heavy metal sources in sediment and waters: a case study in Yangzhong, China. *Environ. Geol. 54*(2), 373-380.

Zhu, H-n., Yuan, X-z., Zeng, G-m., Jiang, M., Liang, J., Zhang C., Yin, J., Huang H-j., Liu, Z-f. & Jiang, H-w. (2012). Ecological risk assessment of heavy metals in sediments of Xiawan Port based on modified potential ecological risk index. *Trans. Nonferrous Met. Soc. China*, *22*, 1470-1477.

Chapter 3

Radioactivity Concentrations and their Radiological Significance in Sediments of the Tema Harbour (Greater Accra, Ghana)

A modified version of this chapter was published as: Botwe, B. O., Schirone, A., Delbono, I., Barsanti, M., Delfanti, R., Kelderman, P., Nyarko, E., & Lens, P. N. L. (2017). Radioactivity concentrations and their radiological significance in sediments of the Tema Harbour (Greater Accra, Ghana). *Journal of Radiation Research and Applied Sciences, 10, 63-71*.

Abstract

Studies on environmental radioactivity in tropical Africa are scarce. Therefore, a baseline study of natural (^{238}U, ^{210}Pb, ^{226}Ra, ^{232}Th, ^{228}Ra, ^{228}Th, ^{40}K) and anthropogenic (^{137}Cs) radionuclides was carried out on Tema Harbour (Greater Accra, Ghana) surface sediments and on their radiological significance. Grab surface sediment samples were collected from 21 stations within the Tema Harbour and their specific activities measured by gamma spectrometry. The mean sediments specific activities (Bq.kg^{-1} dw) were 34 for ^{238}U, 210 for ^{210}Pb, 14 for ^{226}Ra, 30 for ^{232}Th, 29 for ^{228}Ra, 31 for ^{228}Th, 320 for ^{40}K, and 1.5 for ^{137}Cs. Large ^{238}U/^{226}Ra disequilibria were observed in the harbour sediments and a complex dynamics of several mixed sources of sediments within the Tema Harbour can be inferred from the spatial variations in the specific activities. The estimated total absorbed dose rate in air (D), radium equivalent activity (Ra$_{eq}$), external hazard index (H_{ex}), annual gonadal dose equivalent (AGDE) and annual effective dose equivalent (AEDE) indicated no significant radiological risks from the sediment radioactivity concentrations. Application of the Environmental Risk from Ionising Contaminants Assessment and Management tool (ERICA) confirmed that the potential dose rates to biota from the sediment radioactivity concentrations are unlikely to pose appreciable ecological risks. The radioactivity levels are compared with levels reported in sediments from other coastal areas of the world.

3.1. Introduction

Radionuclides constitute an important source of ionising radiation exposure to human and non-human populations (Kam & Bozkurt, 2007; UNSCEAR, 2000), which can cause harmful biological effects such as DNA damage and cancer (Little, 2003; Ravanat et al., 2014; Schmid & Schrader, 2007). Radionuclides such as ^{238}U, ^{210}Pb, ^{226}Ra, ^{232}Th, ^{228}Ra, ^{228}Th and ^{40}K are widely distributed in the environment as a result of their natural occurrence in the Earth's crust or the atmosphere. The human population worldwide receives an average annual radiation dose of 2.4 mSv.y^{-1}, about 80% of which comes from naturally-occurring radionuclides, the remaining part is largely due to artificial sources of which fallout radionuclides account for only 0.4% (UNSCEAR, 2000). Fallout radionuclides such as ^{90}Sr, ^{137}Cs and $^{239+240}$Pu are derived mainly from global nuclear tests conducted between the mid 1940s and the 1980s, as well as from nuclear accidents (Livingston & Povinec, 2000). In addition to their potential ionising effects, radionuclides may be toxic and can undergo bioconcentration and bioaccumulation (Hassona et al., 2008; Sirelkhatim et al., 2008) and adversely impact human and ecosystem health. Assessment of radioactivity in the environment is useful for the protection of human health and the environment from the harmful effects of ionising radiation, and is therefore of great interest (Ulanovsky et al., 2008). Most naturally-occurring and fallout radionuclides can be detected and measured at extremely low concentrations; this makes them excellent tracers for many environmental processes as well as unique dating tools. Thus, they have found a wide range of applications in environmental studies such as isotope hydrogeology (Divine & McDonnel, 2005), water masses circulation (Broecker, 1982), sediment dating and sedimentation (Hernandez, 2016; Mahu et al., 2016; Mabit et al., 2014).

Although the West African coastal environment lacks the presence of nuclear industries, it may be impacted by contaminated areas as a result of ocean and atmospheric dispersal and redistribution. Furthermore, anthropogenic activities in the West African coastal environment such as shipping, offshore oil and gas exploration and production, mining, industrialisation, urbanisation, and agricultural production can potentially add to measured levels of radionuclides in the environment (Al-Trabulsy et al., 2011; El Mamoney & Khater, 2004; Nyarko et al., 2011). In the coastal environment, harbours may be particularly susceptible to anthropogenic influences and their sediments can act as sinks for radionuclides (Sugandhi et al., 2014). Harbours have hence been areas of interest when investigating radionuclide contamination in the coastal environment (Akram et al., 2006; Kumar et al., 2013; Papaefthymiou et al., 2007; Sam et al., 1998; Sugandhi et al., 2014). Sediment contamination by radionuclides of the ^{238}U and ^{232}Th decay-series and ^{40}K is of particular interest from radiological point of view, as they can form the basis of radiological assessments for the human population. The Environmental Risk from Ionising Contaminants Assessment and Management tool (ERICA) developed by the European Commission provides an integrated approach to the assessment and management of environmental risks from ionising radiation (Beresford et al., 2007) and can be applied to assess the potential ecological impact of radionuclide-contaminated environments.

Currently, there is not much information on radioactivity levels in the coastal marine environment of the West African region, including Ghana. In Ghana, the few radioactivity studies in the coastal environment have focused on beach sediments (Amekudzie et al., 2011; Nyarko et al., 2011), estuarine sediments (Mahu et al., 2016) and produce water from offshore oil fields (Kpeglo et al., 2016). Due to rapid growth of urbanisation and industrialisation in

Tema, increasing maritime traffic and infrastructural expansion works at the Tema Port likely intensify chemical contamination in the harbour. Monitoring of such developing coastal areas is, therefore, essential to ensure that these socio-econonic and develomental activities do not adversely affect human health and the proper functioning of coastal aquatic ecosystems. The objective of this study was to assess the specific activities of natural (^{238}U, ^{210}Pb, ^{226}Ra, ^{232}Th, ^{228}Ra, ^{228}Th and ^{40}K) and anthropogenic (^{137}Cs) radionuclides in surface sediments from the Tema Harbour in Ghana and their radiological and radioecological significance.

3.2. Materials and Methods

3.2.1. Description of the study area

The Tema Harbour, situated in the Gulf of Guinea along the Ghana coast at Tema (Fig. 3.1), is semi-enclosed with a water area of 1.7 km^2 within a wider harbour zone of 3.9 km^2. It consists of the Main Harbour, the Fishing Harbour comprising the Inner Fishing Harbour, the Outer Fishing Harbour, and the Canoe Basin. Water depths range from 7.0-11.5 m for the Main Harbour, 7.5-8.5 m for the Fishing Harbour and 3.0-5.0 m for the Canoe Basin. The Main Harbour has 14 berths, a total of 4,580 m breakwater, a shipyard and dry dock for ship repairs, and a 240 m wide access channel. The harbour does not receive riverine inflows, but does receive wastewater from the Tema Township. The Fishing Harbour serves as a landing site for fishing vessels, where repairs and re-fuelling of marine crafts are also carried out. A 2010 feasibility study report by Halcrow Engineers (unpublished) on the Tema Harbour revealed that it is underlain mainly by gneiss rocks composed of feldspar, quartz and micaceous minerals. The Tema Harbour is located in an industrial environment and thus, it is subject to industrial activities.

3.2.2. Sediment sampling

Grab surface sediment samples were collected in November 2013 from 21 stations within the Tema Harbour (Fig. 3.1), excluding rocky areas and berths where ships had docked (e.g. the area between S7 and S14). Fourteen (14) of the sampling stations were located in the Main Harbour (S1-S14), three (3) stations in the Canoe Basin (S15-S17), and two (2) stations each in the Inner Fishing Harbour (S18 and S19) and the Outer Fishing Harbour (S20 and S21). Geographical coordinates of the sampling stations were recorded using a Garmin Global Positioning System (GPS). At each station, the redox potential (E$_h$) was measured *in situ* near the sediment-water interface using a Hanna multi-parameter probe (HI 9829, Hanna Instruments, USA).

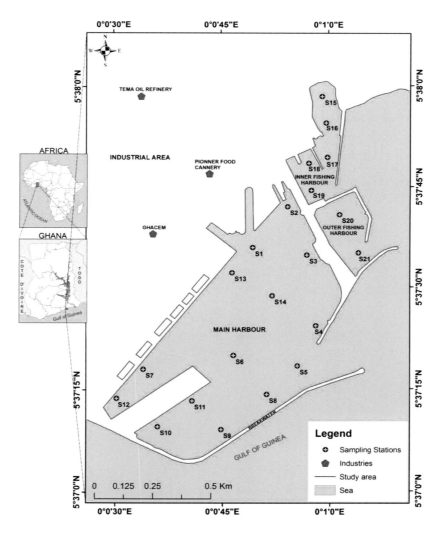

Fig 3.1 Map of Tema Harbour (Greater Accra, Ghana) showing sediment sampling locations

Sediment samples were collected using a 3.5 L Ekman bottom grab sampler. The sampling locations were approximately 200 m apart, covering a wide area and range of water depths to provide representative data on the radionuclide distribution in the harbour. To minimise potential loss of fine particles via leakage of water from the grab, it was ensured that only

77

grabs that arrived firmly closed on the deck were sampled for analysis. In addition, only grabs that were not filled with sediment to the lid were used to assure minimal disturbance of the surface sediments.

About 100 g wet weight portions of surface sediments were taken with a clean plastic spoon into polyamide Rilsan® bags and securely closed. With extremely low potential for diffusion of materials across their surfaces (http://tub-ex.com/products/rilsan/), Rilsan® bags are suitable for collection and storage of sediment samples and they have been used in environmental monitoring programmes (Serigstad et al., 2010). To minimise contamination from the grab, sediments in direct contact with the grab were not used and the sampling spoon was washed with deionised water after each sampling. All sediment samples were kept on ice and transported to the Chemistry Laboratory at the Ghana Atomic Energy Commission (Greater Accra, Ghana) for further analyses.

3.2.3. Sample analyses

For the analyses of radionuclides, wet sediment samples were oven-dried at 50°C till constant weight. Radiometric analyses of sediment samples were conducted at the ENEA S. Teresa laboratory (Italian National Agency for New Technologies, Energy and Sustainable Economic Development, La Spezia, Italy). The sediment samples were ground and placed in 5 g plastic vials of standard geometry, closed and sealed air-tight, and then stored for at least 22 days to ensure secular equilibrium between the parent nuclides and their short-lived daughter nuclides. The sealed samples were then counted for 2-3 days, and activities measured for ^{210}Pb at 46.5 keV, ^{214}Pb at 352 keV, ^{212}Pb at 239 keV, ^{208}Tl at 583 KeV, ^{228}Ac at 338 and 911.0 keV, ^{234}Th at 63.3 and 92.5 keV, ^{40}K at 1460 keV and ^{137}Cs at 662 keV. The measured activities were decay-corrected with respect to the date of sediment sampling (Sirelkhatim et al., 2008) and associated errors were determined from l-sigma counting statistics (Nyarko et al., 2011). The activity of ^{228}Th was obtained from the activity of its daughter ^{212}Pb radionuclide and ^{226}Ra and ^{228}Ra from the activities of ^{214}Pb and ^{228}Ac, respectively. The average activity of ^{228}Ac and ^{212}Pb was used as a proxy for ^{232}Th activity. Since no unsupported ^{234}Th (^{234}Th$_{exc}$) was observed, the supported ^{234}Th activity was used as a proxy for ^{238}U activity.

Samples were analysed using a Gamma spectrometer coupled to an ORTEC low background intrinsic germanium coaxial detector (17.6% absolute efficiency, 1.8 keV nominal resolution at 662 keV ^{137}Cs gamma emission). Prior to the sample radioactivity analyses, the Gamma ray detectors were calibrated for measurement of [40]K using the IAEA-385 Certified Reference Material (CRM), while calibrations for [238]U series radionuclides were performed using the CANMET (Canada Centre for Mineral and Energy Technology) Reference Standard (DL1a), being a U-Th ore in which ^{210}Pb and [226]Ra exist in secular equilibrium. Calibrations for [232]Th series radionuclides were also performed using the IAEA Reference Standard (RGTh-1) prepared by CANMET, while calibration for ^{137}Cs was performed using the Eckert & Ziegler Analytics Reference Standard (QCYA48). Quality of all results was routinely checked by analysing IAEA-300 and IAEA-315 Reference Materials as well as detector blanks (empty sample containers) processed in a similar way as the actual samples. Correction of measured activities for self-adsorption effects was done based on measurements of the attenuation of a known [210]Pb Gamma source by the samples.

3.2.4. Radiological risk assessment

Dredged radioactive-contaminated harbour sediments may be disposed of on land or used for other purposes such as building, which can potentially result in human exposure to ionising radiations and cause radiological effects. For human protection and appropriate handling of radioactive-contaminated sediments, it is essential to characterise the associated potential radiological risks. Five radiological hazard indices were estimated following established formulae to characterise the potential radiation dose to humans resulting from exposure to sediment radioactivity, *viz.* (1) total absorbed dose rate in air (D), (2) radium equivalent activity (Ra_{eq}), (3) external hazard index (H_{ex}), (4) annual gonadal dose equivalent (AGDE), and (5) annual effective dose equivalent (AEDE).

The D expresses the rate of exposure to gamma radiation in air at 1m above the ground due to the activities of [226]Ra, [232]Th and [40]K in the sediment samples. Ra_{eq} is a weighted sum of the specific activity of [226]Ra, [232]Th and [40]K in a sediment sample, which allows comparison with their individual [226]Ra, [232]Th and [40]K specific activities (Sugandhi et al., 2014). H_{ex} is a measure of the indoor radiation dose rate associated with external gamma radiation exposure from natural radionuclides in building materials and it is important when considering the suitability of sediments as building materials (Xinwei et al., 2006). For human health safety, the value of H_{ex} must not exceed 1.0 (Kurnaz et al., 2007; Xinwei et al., 2006). Owing to their relatively higher sensitivity to ionising radiation compared to other organs of the body, the gonads are considered to be at a high risk of radiation exposure and are therefore of great interest in radiological assessment (Kurnaz et al., 2007). The AGDE estimates the potential radiation dose that the gonads may receive from ^{226}Ra, ^{232}Th and ^{40}K. It is also usual to convert D to AEDE to assess the dose rate to an individual from outdoor gamma radiation over a period of one year.

The D and Ra_{eq} were calculated using Eq. 1 and 2, respectively, following (El Mamoney & Khater, 2004):

$$D \ (nGy.h^{-1}) = 0.462A_{Ra} + 0.604A_{Th} + 0.0417A_{K} \qquad (1)$$

$$Ra_{eq} \ (Bq.kg^{-1}) = A_{Ra} + 1.43A_{Th} + 0.077A_{K} \qquad (2)$$

H_{ex} was calculated using Eq. 3 following (Xinwei et al., 2006):

$$H_{ex} = (A_{Ra}/370) + (A_{Th}/259) + (A_{K}/4810) \qquad (3)$$

AGDE and AEDE were calculated using Eq. 4 and 5, respectively, following (Kurnaz et al., 2007).

$$AGDE \ (\mu Sv.y^{-1}) = 3.09A_{Ra} + 4.18A_{Th} + 0.314A_{K} \qquad (4)$$

$$AEDE \ (\mu Sv.y^{-1}) = D \ (nGy.h^{-1}) \ x \ 8760 \ h \ x \ 0.2 \ x \ 0.7 \ Sv.Gy.y^{-1} \ x \ 10^{-3} \qquad (5)$$

3.2.5. Radioecological risk assessment

Radioactivity contamination in sediments may put aquatic organisms at risk of ionising radiation effects. Therefore, the ERICA tool (version 1.2) was applied to assess the potential dose rates to organisms in the harbour. A detailed description of the ERICA tool can be found in literature (Beresford et al., 2007; Brown et al., 2008; Larsson, 2008). The tool is based on data gathered from extensive radioecological and dosimetric studies and uses generalised ecosystem representations, also referred to as reference organisms (Beresford et al., 2007). In this context, a reference organism is defined as *'a series of entities that provide a basis for the estimation of radiation dose rate to a range of organisms which are typical, or representative, of a contaminated environment'* (Beresford et al., 2007).

In this study, the ERICA tool was used to estimate radioactivities in ten (10) reference organisms captured in the ERICA database, *viz. phytoplankton, zooplankton, macroalgae, pelagic/benthic fishes, mammals, crustaceans, mollusc-bivalves, sea anemones,* and *polychaete worms.* In estimating the radioactivities in the selected reference organisms, the highest measured specific activities in the sediments (Bq.kg^{-1} dry wt.) were used as input data to represent "worst case scenarios". Default concentration ratios for the reference organisms in ERICA were then applied. Since, by default, ^{40}K was not included in the ERICA database, this isotope was also not considered here.

Furthermore, a Tier 2 ERICA assessment was conducted to estimate the total dose rates to biota, applying a default uncertainty factor of 3.0 in the ERICA tool to ensure there will be less than 5% probability of modelled dose rates exceeding the screening dose rate. Since the magnitude of biological effects varies with different types of ionising radiation (Schmid & Schrader, 2007), default ERICA weighting factors of 10.0 for alpha, 1.0 for beta/gamma, and 3.0 for low beta radiation were applied to give appropriate weights to the dose rates. As dose rate to biota is a function of the duration of exposure, a default ERICA occupancy factor of 1.0 (i.e., fraction of time that the organism spends at a specified location in its habitat) was assigned to each reference organism, assuming they spend 100% of the time at their specified locations (i.e. in the water column, on the sediment surface or inside the sediment).

3.3.   Results and discussion

3.3.1. Radioactivity levels in the Tema Harbour sediments

The specific activities of ^{238}U, ^{210}Pb, ^{226}Ra, ^{228}Ra, ^{228}Th, ^{232}Th, ^{40}K and ^{137}Cs in the surface sediment samples are shown in Table 3.1. The specific activities of ^{40}K were relatively higher than those of the other radionuclides, ranging from 250 to 570 Bq.kg^{-1} with a mean of 320 Bq.kg^{-1}. Doyi et al. (2013) reported higher levels of ^{40}K and ^{232}Th for rocks and ore from mines in the Upper East Region of Ghana in the ranges of 950-2800 and 81-880 Bq.kg^{-1}, respectively, with a higher mean ^{238}U level of 66 (± 8) Bq.kg^{-1}. The levels of ^{210}Pb (except at station S16) were markedly higher than those of ^{238}U, ^{226}Ra, ^{232}Th, ^{228}Ra and ^{228}Th. Their mean value, 210 (± 10) Bq.kg^{-1} (~200 Bq.kg^{-1} for unsupported ^{210}Pb) is to some extent higher, but comparable to the values of unsupported ^{210}Pb found by Mahu et al. (2016) in the upper layers of sediment cores sampled in the Amisa (103 Bq.kg^{-1}), Sakumo II (157 Bq.kg^{-1}) and Volta (123 Bq.kg^{-1}) estuaries in Ghana, respectively.

In coastal marine sediments, an excess of ^{228}Th above that supported by the parents ^{232}Th or ^{228}Ra has often been reported, and used for radiometric dating (Koide et al., 1973). In the Tema Harbour sediments, the specific activities of ^{232}Th, ^{228}Ra and ^{228}Th were comparable within the involved uncertainties. Specific activities of ^{137}Cs in the sediments were markedly low, varying from <0.3 to 2.3 Bq.kg^{-1} with a mean of 1.5 Bq.kg^{-1}. These values are comparable to those reported by Mahu et al. (2016) for the surface layers of estuarine sediments from Ghana (in the range 0-7 Bq.kg^{-1}). The low ^{137}Cs levels suggest low atmospheric fallout in the study area, coupled with natural decay following its deposition (Livingston & Povinec, 2000; Pfitzner et al., 2004) or remobilisation from sediment into seawater (Sugandhi et al., 2014). Thus, Junge et al. (2013) have reported ^{137}Cs concentrations in the range 0.5-6.5 Bq.kg^{-1} in farmland soils from Nigeria.

The levels of ^{210}Pb (20 ± 5 Bq.kg^{-1}) and ^{137}Cs (< 0.3 Bq.kg^{-1}) in the shallow station of the Canoe Basin (S16) were very low compared to the levels found at the remaining stations. Several mechanisms are involved in the distribution of radionuclides in the harbour such as hydrodynamics, waves, tides, and vessel movements and dredging. The Canoe Basin is prone to receive sediments from mixed sources including sand bars bordering one of its sides. In May 2013, maintenance dredging was carried out in the Canoe Basin to ensure safer navigation and increase berthing capacity for the operation of canoes. Dredging can potentially remove contaminated sediments while influx of sand may cause dilution, and may partly account for the low radioactivity levels of ^{210}Pb and ^{137}Cs at S16. In the Main Harbour, the ^{210}Pb levels at stations S7 (130 ± 10 Bq.kg^{-1}), S11 (110 ± 10 Bq.kg^{-1}) and S13 (130 ± 10 Bq.kg^{-1}) were relatively lower than those at the other stations. This may partly be due to a dilution effect of the discharges of industrial wastewater and sand channelled into the harbour at S7, which can also affect neighbouring areas such as stations S13 and S11. On the contrary, Table 3.1 shows that these same three stations (S7, S11 and S13) recorded relatively higher levels of ^{40}K, possibly due to organic enrichment in sediments as a result of the waste water discharges.

The ratios of $^{238}U/^{226}Ra$, $^{238}U/^{210}Pb$ and $^{210}Pb/^{226}Ra$ were in the ranges of 1.4-3.5, 0.1-0.4 and 1.3-22.4, respectively. The departures of these ratios from 1.0 indicate disequilibria, a general characteristic of marine surface sediments previously observed by others (Chen & Huh, 1999; Koide et al., 1973), which may be caused by anthropogenic influences (Al-Trabulsy et al., 2011; El Mamoney & Khater, 2004; Nyarko et al., 2011; UNSCEAR, 2000).

Table 3.1 Radioactivity concentrations (Bq.kg^{-1} dry wt.) in Tema Harbour surface sediments and water depth and E$_h$ at the sampling stations

SS	WD (m)	E$_h$ (mV)	^{238}U series radionuclides			^{232}Th series radionuclides			^{40}K	^{137}Cs
			^{238}U	^{210}Pb	^{226}Ra	$^{232}Th^*$	^{228}Ra	^{228}Th		
S1	10	-96	30 ± 6	230 ± 10	12 ± 2	21 ± 3	18 ± 2	24 ± 1	250 ± 10	1.6 ± 0.4
S2	7	-128	39 ± 5	200 ± 10	11 ± 1	25 ± 3	24 ± 2	26 ± 1	260 ± 10	1.7 ± 0.5
S3	8.5	-100	43 ± 5	250 ± 10	13 ± 1	26 ± 4	26 ± 3	25 ± 1	270 ± 10	2.0 ± 0.5
S4	8	-70	40 ± 4	260 ± 10	18 ± 2	32 ± 4	30 ± 3	33 ± 1	350 ± 10	1.1 ± 0.5
S5	8	-65	28 ± 5	250 ± 10	14 ± 2	21 ± 3	19 ± 2	22 ± 1	250 ± 10	1.1 ± 0.5
S6	8	-94	28 ± 4	250 ± 10	17 ± 2	70 ± 4	67 ± 3	72 ± 1	330 ± 10	1.3 ± 0.5
S7	8	-90	28 ± 4	130 ± 10	13 ± 1	37 ± 3	34 ± 2	40 ± 1	460 ± 20	0.9 ± 0.5
S8	9	-80	28 ± 4	210 ± 10	13 ± 1	19 ± 3	18 ± 2	20 ± 1	270 ± 10	1.2 ± 0.4
S9	8.5	-80	35 ± 4	290 ± 10	16 ± 2	27 ± 3	24 ± 2	29 ± 1	290 ± 10	2.3 ± 0.5
S10	9.5	-78	32 ± 4	230 ± 10	15 ± 1	26 ± 3	24 ± 2	28 ± 1	350 ± 10	1.1 ± 0.4
S11	8.5	70	21 ± 4	110 ± 10	10 ± 1	38 ± 3	36 ± 2	39 ± 1	570 ± 20	0.5 ± 0.3
S13	9	50	47 ± 7	130 ± 10	14 ± 1	44 ± 4	44 ± 3	43 ± 1	390 ± 20	2.2 ± 0.5
S14	7.5	70	36 ± 3	240 ± 10	17 ± 1	29 ± 3	27 ± 2	31 ± 1	330 ± 10	1.1 ± 0.4
S15	10	95	42 ± 4	280 ± 10	16 ± 1	28 ± 3	25 ± 2	31 ± 1	330 ± 10	1.4 ± 0.4
S16	3	400	24 ± 4	20 ± 5	16 ± 1	37 ± 3	36 ± 2	38 ± 1	330 ± 10	< 0.3
S17	4	-210	26 ± 5	230 ± 10	13 ± 2	34 ± 3	35 ± 2	32 ± 1	310 ± 10	1.9 ± 0.5
S18	5	-140	32 ± 4	230 ± 10	13 ± 2	25 ± 3	21 ± 2	28 ± 1	310 ± 10	1.5 ± 0.4
S19	8	-115	42 ± 5	310 ± 20	14 ± 1	23 ± 3	22 ± 2	23 ± 1	270 ± 10	1.9 ± 0.4
S20	8	-110	35 ± 6	240 ± 10	11 ± 2	24 ± 4	23 ± 3	25 ± 1	300 ± 10	1.6 ± 0.5
S21	8	-100	27 ± 4	170 ± 10	20 ± 1	26 ± 3	24 ± 2	27 ± 1	300 ± 10	1.4 ± 0.4
S22	8	-90	41 ± 5	170 ± 10	18 ± 1	28 ± 3	27 ± 2	29 ± 1	300 ± 10	1.5 ± 0.5
Mean	-	-	34 ± 5	210 ± 10	14 ± 1	30 ± 3	29 ± 2	31 ± 1	320 ± 10	1.5 ± 0.5

SS = sampling station; Errors are 1 standard deviation from counting statistics; $^{*232}Th$ activities are estimated from grand-daughters activities

The origin of the excess in ^{210}Pb found in sediments is well known, and it is the basis of the ^{210}Pb-based radiometric method for dating recent sediments (Mabit et al, 2014). On the contrary, the observed disequilibria between ^{238}U (derived from ^{234}Th activity, see Section 3.2.3) and ^{226}Ra may be ascribed to differences in geochemical behaviour of the radionuclides (Koide et al., 1973). ^{226}Ra and ^{238}U are known for their different sorption and mobility characteristics; ^{226}Ra and ^{238}U are less particle-reactive than Th isotopes with greater tendency to diffuse from sediments into the surrounding seawater (Chen & Huh, 1999; Sirelkhatim et al., 2008; Sugandhi et al., 2014). The higher levels of ^{238}U relative to ^{226}Ra in the sediments may be due to diffusion and loss of Ra from sediment owing to its higher solubility in seawater or higher leachability of ^{226}Ra from the harbour sediments than ^{238}U-^{234}Th. At the time of sampling, the bottom water was anoxic, having E_h values in the range of -50 to -400 mV (Table 3.1), which may enhance immobilisation and precipitation of ^{238}U in the sediments (El Mamoney & Khater, 2004). Mohamed et al. (2010) reported ^{228}Ra/^{226}Ra activity ratios in the range 1.2-2.9 with ^{238}U/^{226}Ra activity ratios in the range 1.1-5.6 for southern South China Sea surface sediments, which compare well with the ranges 1.4-3.9 and 1.6-3.5 found in this work for ^{228}Ra/^{226}Ra and ^{238}U/^{226}Ra activity ratios, respectively (Table 3.1).

The mean specific activity levels in the Tema Harbour sediments and levels reported for other areas are shown in Table 3.2. With the exception of Labadi that recorded higher levels of ^{226}Ra and ^{232}Th, the mean specific activity levels in the Tema Harbour sediments were generally higher than those reported in sediments from other areas, i.e. Chorkor, James Town, Nungua, Kokrobite, Teshie and Weija of the Greater Accra coast of Ghana (Amekudzie et al., 2011). Nyarko et al. (2011) also reported lower specific activities of ^{210}Pb in beach sand samples along the Ghana coast ranging from 1.6 up to 4.5 Bq.kg^{-1} and ^{137}Cs concentrations below the detection limit of 0.4 Bq.kg^{-1}. The mean specific activity levels in the Tema Harbour sediments were also higher than the levels reported for the Mediterranean coast of Egypt (Higgy, 2000), but were comparable to levels reported for the Saudi coast of the Gulf of Aqaba, except for their higher ^{40}K (Al-Trabulsy et al., 2011); the Red Sea coast of Egypt (El Mamoney & Khater, 2004), Port Sudan and Sawakin Harbour of Sudan (Sam et al., 1998), Patras Harbour, Greece (Papaefthymiou et al., 2007), Mumbai Harbour, India (Sugandhi et al., 2014), and the Karachi Harbour, Pakistan (Sugandhi et al., 2014). The mean specific activity levels of ^{238}U, ^{232}Th, ^{40}K and ^{137}Cs in the Tema Harbour sediments were, however, much lower than those found in polluted sediments from the Caspian Sea coast of Iran (Abdi et al., 2009). Mohamed et al. (2010) found comparatively higher levels of ^{226}Ra, ^{228}Ra, ^{238}U and

Table 3.2 Specific activity levels in Tema Harbour surface sediments and levels reported in sediments from other areas of the world

Sampling Station	238U series radionuclides (Bq.kg-1)			232Th series radionuclides (Bq.kg-1)			40K (Bq.kg-1)	137Cs (Bq.kg-1)	Reference
	238U	210Pb	226Ra	232Th	228Ra	228Th			
Ghana coast									
Chorkor	-	-	1.42	1.49	-	-	21.31	-	Amekudzie et al. (2011)
James Town	-	-	0.82	1.04	-	-	14.67	-	"
Labadi	-	-	140.8	732.6	-	-	43.97	-	"
Nungua	-	-	4.05	8.64	-	-	41.17	-	"
Kokrobite	-	-	3.74	6.63	-	-	17.76	-	"
Teshie	-	-	2.85	9.66	-	-	61.01	-	"
Weija	-	-	0.62	0.17	-	-	8.6	-	"
Tema Habour	34	210	14	30	29	31	325	1.5	This study
Other areas									
Port Sudan, Sudan	-	-	11.05	-	10.35	-	311	7.02	Sam et al. (1998)
Sawakin Harbour, Sudan	-	-	12.61	-	6.18	-	192	4.51	"
Patras Harbour, Greece	-	-	22.6	24.5	-	-	497	3.1	Papaefthymiou et al. (2007)
Mumbai Harbour, India	-	-	10.6	-	12.7	-	436	21.6	Sugandhi et al. (2014)
Karachi Harbour, Pakistan	-	-	23.9	-	23.5	-	527	< 1.3	Akram et al. (2006)
Saudi coast, Gulf of Aqaba	16.9	-	11	22	19	-	641	3.5	Al-Trabulsy et al. (2011)
Red Sea coast, Egypt	25.5	26	24.6	-	-	-	427.5	-	El Mamoney and Khater (2004)
Mediterranean coast, Egypt	8.8	-	5	2.1	-	-	46	-	Higgy (2000)
Caspian Sea coast, Iran	177	-	-	117	-	-	1085	131	Abdi et al. (2009)
Malaysia coast, South China Sea	45.9	-	27.7	73.3	66.2	-	-	-	Mohamed et al. (2010)
World average	35[a]	25[b]	25[b,c,d,e]	25[c,e]	-	-	373[d,e]	-	[a]SureshGandhi et al. (2014); [b]Nyarko et al. (2011); [c]Qureshi et al. (2014); [d]Kurnaz et al. (2007); [e]Sugandhi et al. (2014)

[232]Th for surface sediments from the Malaysia coast of the southern South China Sea. Generally, the mean specific activity levels in the Tema Harbour sediments were comparable to the reported worldwide averages (Nyarko et al., 2011; Papaefthymiou et al., 2007; Qureshi et al., 2014; SureshGandhi et al., 2014), except for ^{210}Pb which was almost an order of magnitude higher than its grandparent [226]Ra, due to the presence of unsupported [210]Pb.

3.3.2. Radiological significance of Tema Harbour sediment specific activity levels

The calculated hazard indices for the Tema Harbour sediments and hazard indices reported for sediments from other areas of the world are presented in Table 3.3. The ranges (means) of D, Ra_{eq}, H_{ex}, AGDE, and AEDE for the Tema Harbour sediments were 29-64 nGy.h^{-1} (39 nGy.h^{-1}), 61-142 Bq.kg^{-1} (82.7 Bq.kg^{-1}), 0.2-0.4 (0.2), 203-447 µSv.y^{-1} (273 µSv.y^{-1}), and 35-78 µSv.y^{-1} (47 µSv.y^{-1}), respectively. These mean D, Ra_{eq}, H_{ex}, AGDE, and AEDE values were below the recommended values, although there were a few areas where the measured Ra_{eq} (i.e. S4, S6, S7, S11, S13 and S16), ADGE (i.e. S6, S7, S11, S13 and S16) and AEDE (i.e. S6) exceeded the recommended values. Thus, the radioactivity levels in the harbour sediments are generally of little radiological concern for human health.

Compared to this study, Amekudzie et al. (2011) reported higher mean D (77 nGy.h^{-1}) and H_{ex} (0.5), but lower mean Ra_{eq} (9 Bq.kg^{-1}) values for sediments from Chorkor, James Town, Nungua, Kokrobite, Teshie and Weija along the Greater Accra coast of Ghana (Table 3.3). Comparable mean values of D (42 nGy.h^{-1}) and Ra_{eq} (101 Bq.kg^{-1}) were reported for the Red Sea coast of Egypt by El Mamoney & Khater (2004), while a lower mean D value of 5.5 nGy.h^{-1} was reported for the Mediterranean coast of Egypt by Higgy (2000). Higher mean values of D (63 nGy.h^{-1}), Ra_{eq} (176 Bq.kg^{-1}) and H_{ex} (0.5) were reported for sediments from the Caspian Sea coast by Abdi et al. (2009), while a lower mean H_{ex} value of 0.1 was reported for the Saudi coast of the Gulf of Aqaba by Al-Trabulsy et al. (2011).

Table 3.3 Calculated D, Ra_{eq}, H_{ex}, AGDE and AEDE for Tema Harbour surface
sediments and sediments from other parts of the world as well as
recommended values

	D (nGy.h^{-1})	Ra_{eq} (Bq.kg^{-1})	H_{ex}	AGDE (μSv.y^{-1})	AEDE (μSv.y^{-1})	Reference
Tema Harbour						This study
S1	29	61	0.2	203	35	
S2	31	67	0.2	220	38	
S3	33	70	0.2	232	40	
S4	42	90	0.2	297	51	
S5	29	63	0.2	208	36	
S6	64	142	0.4	447	78	
S7	48	101	0.3	339	58	
S8	29	61	0.2	204	35	
S9	36	76	0.2	251	44	
S10	37	79	0.2	265	46	
S11	51	108	0.3	367	63	
S13	49	106	0.3	348	60	
S14	39	84	0.2	277	48	
S15	38	82	0.2	270	47	
S16	44	94	0.3	308	53	
S17	39	85	0.2	278	48	
S18	34	72	0.2	240	41	
S19	31	67	0.2	222	38	
S20	32	68	0.2	229	39	
S21	37	80	0.2	263	46	
S22	38	81	0.2	267	46	
Mean	39	83	0.2	273	47	
Mean values for other areas						
Caspian Sea coast, Iran	63	176	0.5	-	-	Abdi et al. (2009)
Red Sea coast, Egypt	42	101	-	-	-	El Mamoney and Khater (2004)
Saudi coast, Gulf of Aqaba	-	-	0.1	-	-	Al-Trabulsy et al. (2011)
Mediterranean coast, Egypt	5.5	-	-	-	-	Higgy (2000)
Ghana coast	77	9	0.5	-	0.1	Amekudzie et al. (2011)
Recommended value	55[a,b]	89[a]	1.0[c]	300[c]	70[c]	

D (total absorbed dose rate in air); Ra_{eq} (radium equivalent activity); H_{ex} (external hazard

index); AGDE (annual gonadal dose equivalent); AEDE (annual effective dose equivalent);

[a]Kurnaz et al. (2007); [b]Abdi et al. (2009); [c]Xinwei et al. (2006); - : not reported

3.3.3. Radioecological significance of Tema Harbour sediment radioactivity

Table 3.4 presents the estimated specific activity levels in the selected reference organisms. Generally, biota exhibited potentially higher accumulation levels of ^{210}Pb (1.18-570 Bq.kg^{-1}) relative to the other radionuclides (0-0.05 for ^{137}Cs; 0.82-14.4 for Ra; 0.01- 6.60 for Th, and 0.06-17.5 Bq.kg^{-1}for ^{238}U), possibly reflecting the relatively higher levels of ^{210}Pb in the sediments. Thus, sediments may be an important source of ^{210}Pb exposure to biota. The levels of ^{210}Pb were markedly high in phytoplankton, exceeding the levels in the sediments. This is an indication of a high potential for ^{210}Pb bioaccumulation by phytoplankton as has been reported by Hassona et al. (2008). Overall, phytoplankton exhibited the highest potential for radionuclide exposure from sediment, suggesting that it could be a good bioindicator for monitoring of radionuclide contamination in the Tema Harbour.

Apart from the ^{210}Pb levels in phytoplankton, the specific levels in all the reference organisms were generally low compared to the levels in the sediments. This indicates that the radionuclides associate mainly with the sediments, despite their potential for bioaccumulation (Hassona et al., 2008; Sirelkhatim et al., 2008). This supports the view that sediments are the major sinks for radionuclides in aquatic ecosystems (Sugandhi et al., 2014). The very low specific activity levels of Th and ^{137}Cs in biota suggest that sediments may not be a major exposure pathway of these radionuclides to biota in the harbour.

The ERICA-derived dose rates to biota are presented in Table 3.5. It shows that phytoplankton could potentially receive higher dose rates from sediment radioactivity than the other reference organisms, which corresponds to its higher bioaccumulation potential for radionuclides (Table 3.4). The total dose rate to phytoplankton slightly exceeded the typical maximum value in the ERICA database (see Table 3.5). Whereas the total dose rate to pelagic fishes was similar to the typical minimum, that of crustaceans slightly exceeded the typical minimum value, but fell below the typical average value. In the case of zooplankton, macroalgae, benthic fishes, mammal, mollusc-bivalves and polychaete worms, the total dose rates fell below their ERICA typical minimum values (Table 3.5). The total dose rates to all the reference organisms fell below the screening dose rate of 400 µGy.h^{-1}, proposed by the International Atomic Energy Agency (IAEA) and the United Nations Scientific Committee on the Effects of Atomic Radiation (UNSCEAR) as the dose rate below which harmful effects are unlikely to occur in organisms.

Table 3.4 Highest specific activities in Tema Harbour sediments (Bq.kg^{-1} d.w) and ERICA-derived specific activities in reference organisms (Bq.kg^{-1} f.w)

					Reference organisms					
	Sediment	Phyto-plankton	Zoo-plankton	Macro-algae	Pelagic/Benthic fishes	Mammals	Crustaceans	Mollusc-bivalves	Sea anemones	Polychaete worms
^{137}Cs	2.3	0	0.03	0.02	0.02	0.05	0.01	0.01	0.05	0.04
*Ra	67	14.4	1.02	1.13	1.76	2.06	1.08	0.82	1.76	1.76
^{210}Pb	314	570	20	1.18	38.9	22.4	24.8	7.43	38.9	47.2
*Th	72	6.60	0.06	0.04	0.01	0.02	0.35	0.02	0.02	0.02
^{238}U	47	3.85	0.07	1.47	0.16	0.16	0.06	0.57	17.5	17.5

*Ra is for both ^{226}Ra and ^{228}Ra; *Th for both ^{232}Th and ^{228}Th

Table 3.5 ERICA-derived total dose rates to reference organisms and typical maximum, minimum, and average values (uGy.h^{-1}) in the ERICA database

Reference Organism	Total dose rate	Typical average dose rate	Typical minimum dose rate	Typical maximum dose rate
Phytoplankton	2.11	0.38	0.13	2.00
Zooplankton	0.06	0.94	0.25	5.20
Pelagic fish	0.09	0.42	0.08	3.70
Mammal	0.10	1	0.23	5.80
Macroalgae	0.13	0.87	0.52	1.40
Benthic fish	0.12	0.58	0.24	1.20
Crustacean	0.16	0.59	0.12	1.90
Mollusc-bivalve	0.09	2	0.98	5.60
Sea anemones	0.54	4.20	1.90	8.80
Polychaete worm	0.58	1.6	0.94	2.5

It is worthy to note that the ERICA tool was used in this study as a screening tool to provide estimates of the radioactivity levels in biota and associated total dose rates rather than their accurate prediction. Furthermore, the estimated total dose rates to biota may be underestimated since they were based on the specific activities of only ^{238}U, ^{210}Pb, ^{226}Ra, ^{228}Ra, ^{228}Th, ^{232}Th and ^{137}Cs. Measurement of radioactivity levels of ^{238}U, ^{210}Pb, ^{226}Ra, ^{228}Ra, ^{228}Th, ^{232}Th and ^{137}Cs in different environmental matrices such as sediments, water and biota from the harbour will be important to validate the predictive ability of the ERICA modelling tool for the Tema Harbour ecosystem and is, therefore, recommended for future studies.

3.4. Conclusions

The radioactivity levels of ^{238}U, ^{210}Pb, ^{226}Ra, ^{232}Th, ^{228}Ra, ^{228}Th, ^{40}K and ^{137}Cs in surface sediment from the Tema Harbour in Ghana have been assessed for the first time. The specific activities of ^{40}K and ^{210}Pb were relatively higher than the levels of the other radionuclides in the Tema Harbour sediments. Large disequilibria between ^{238}U and ^{226}Ra were found, attributable to the dynamics of the sediments in the harbour. Apart from ^{210}Pb, the specific activities of the natural radionuclides in the Tema Harbour sediments were comparable to worldwide average values. Evaluation of total absorbed dose rate in air (D), radium equivalent activity (Ra_{eq}), external hazard index (H_{ex}), annual gonadal dose equivalent (AGDE) and annual effective dose equivalent (AEDE) indicate that the potential dose rates to human from the sediment radioactivity levels may not present significant risks to human health. Moreover, the potential dose rates to biota derived from the ERICA assessment indicate low ecological risks associated with the radioactivity levels in the harbour sediments. This study provides baseline information on radioactivity levels in the Tema Harbour sediments for comparison to future monitoring studies.

References

Abdi, M.R., Hassanzadeh, S., Kamali, M., & Raji, H.R. (2009). ^{238}U, ^{232}Th, ^{40}K and ^{137}Cs activity concentrations along the southern coast of the Caspian Sea, Iran. *Marine Pollution Bulletin, 58*(5), 658-662.

Akram, M., Qureshi, R., Ahmad, N., Solaija, T., Mashiatullah, A., Afzal, M., Faruq, M.U., & Zeb, L. (2006). Concentration of natural and artificial radionuclides in bottom sediments of Karachi Harbour/Manora Channel, Pakistan Coast (Arabian Sea). *Journal of the Chemical Society of Pakistan, 28*(3), 306-312.

Al-Trabulsy, H., Khater, A., & Habbani, F. (2011). Radioactivity levels and radiological hazard indices at the Saudi coastline of the Gulf of Aqaba. *Radiation Physics and Chemistry, 80*(3), 343-348.

Amekudzie, A., Emi-Reynolds, G., Faanu, A., Darko, E., Awudu, A., Adukpo, O., Quaye, L.A.N., Kpordzro, R., Agyeman, B., & Ibrahim, A. (2011). Natural radioactivity concentrations and dose assessment in shore sediments along the coast of Greater Accra, Ghana. *World Applied Sciences Journal, 13*(11), 2338-2343.

Beresford, N., Brown, J., Copplestone, D., Garnier-Laplace, J., Howard, B., Larsson, C.-M., Oughton, D., Pröhl, G., & Zinger, I. (2007). D-ERICA: An integrated approach to the assessment and management of environmental risk from ionising radiation. Description of purpose, methodology and application.

Broecker, W. (1982). Geochemical tracers and ocean circulation. *Lamont-Doherty Geological Observatory, Palisades, NY*, 434-450.

Brown, J., Alfonso, B., Avila, R., Beresford, N.A., Copplestone, D., Pröhl, G., & Ulanovsky, A. (2008). The ERICA tool. *Journal of Environmental Radioactivity, 99*(9), 1371-1383.

Chen, H.Y., & Huh, C.A. (1999). ^{232}Th–^{228}Ra–^{228}Th disequilibrium in East China Sea sediments. *Journal of Environmental Radioactivity*, *42*(1), 93-100.

Divine, C.E., & McDonnell, J.J. (2005). The future of applied tracers in hydrogeology. *Hydrogeology Journal, 13*(1), 255-258.

Doyi, I., Oppon, O., Glover, E., Gbeddy, G., & Kokroko, W. (2013). Assessment of occupational radiation exposure in underground artisanal gold mines in Tongo, Upper East Region of Ghana. *Journal of Environmental Radioactivity, 126*, 77-82.

El Mamoney, M., & Khater, A.E. (2004). Environmental characterization and radio-ecological impacts of non-nuclear industries on the Red Sea coast. *Journal of Environmental Radioactivity, 73*(2), 151-168.

Hassona, R.K., Sam, A.K., Osman, O.I., Sirelkhatim, D.A., & LaRosa, J. (2008). Assessment of committed effective dose due to consumption of Red Sea coral reef fishes collected from the local market (Sudan). *Science of the Total Environment, 393*(2–3), 214-218.

Hernández, J.-M.A. (2016). A ^{210}Pb-based chronological model for recent sediments with random entries of mass and activities: Model development. *Journal of Environmental Radioactivity, 151*, 64-74.

Higgy, R. (2000). Natural radionuclides and plutonium isotopes in soil and shore sediments on Alexandria Mediterranean Sea coast of Egypt. *Radiochimica Acta, 88*(1), 47-54.

Junge, B., Mabit, L., Dercon, G., Walling, D., Abaidoo, R., Chikoye, D., & Stahr, K. (2010). First use of the 137 Cs technique in Nigeria for estimating medium-term soil redistribution rates on cultivated farmland. *Soil and Tillage Research, 110*(2), 211-220.

Kam, E., & Bozkurt, A. (2007). Environmental radioactivity measurements in Kastamonu region of northern Turkey. *Applied Radiation and Isotopes, 65*(4), 440-444.

Koide, M., Bruland, K.W., & Goldberg, E.D. (1973). Th-228/Th-232 and Pb-210 geochronologies in marine and lake sediments. *Geochimica et Cosmochimica Acta, 37*(5), 1171-1187.

Kpeglo, D., Mantero, J., Darko, E., Emi-Reynolds, G., Faanu, A., Manjón, G., Vioque, I., Akaho, E.H.K., & Garcia-Tenorio, R. (2016). Radiochemical characterization of produced water from two production offshore oilfields in Ghana. *Journal of Environmental Radioactivity, 152*, 35-45.

Kumar, A., Karpe, R., Rout, S., Joshi, V., Singhal, R.K., & Ravi, P.M. (2013). Spatial distribution and accumulation of ^{226}Ra, ^{228}Ra, ^{40}K and ^{137}Cs in bottom sediments of

Mumbai Harbour Bay. *Journal of Radioanalytical and Nuclear Chemistry, 295*(2), 835-839.

Kurnaz, A., Küçükömeroğlu, B., Keser, R., Okumusoglu, N., Korkmaz, F., Karahan, G., & Çevik, U. (2007). Determination of radioactivity levels and hazards of soil and sediment samples in Fırtına Valley (Rize, Turkey). *Applied Radiation and Isotopes, 65*(11), 1281-1289.

Larsson, C.-M. (2008). An overview of the ERICA Integrated Approach to the assessment and management of environmental risks from ionising contaminants. *Journal of Environmental Radioactivity, 99*(9), 1364-1370.

Little, M. (2003). Risks associated with ionizing radiation Environmental pollution and health. *British Medical Bulletin, 68*(1), 259-275.

Livingston, H.D., & Povinec, P.P. (2000). Anthropogenic marine radioactivity. *Ocean & Coastal Management, 43*(8), 689-712.

Mabit, L., Benmansour, M., Abril, J., Walling, D., Meusburger, K., Iurian, A.R., Bernard, C., Tarján, S., Owens, P.N., Blake, W.H., & Alewell, C. (2014). Fallout ^{210}Pb as a soil and sediment tracer in catchment sediment budget investigations: a review. *Earth-Science Reviews, 138*, 335-351.

Mahu, E., Nyarko, E., Hulme, S., Swarzenski, P., Asiedu, D. K., & Coale, K. H. (2016). Geochronology and historical deposition of trace metals in three tropical estuaries in the Gulf of Guinea. *Estuarine, Coastal and Shelf Science, 177*, 31-40.

Mohamed, C. A. R., Mahmood, Z. U. W., Ahmad, Z., & Ishak, A. K. (2010). Enrichment of natural radium isotopes in the southern South China Sea surface sediments. *Coastal Marine Science, 34*(1), 165-171.

Nyarko, E., Botwe, B., Ansong, J., Delfanti, R., Barsanti, M., Schirone, A., & Delbono, I. (2011). Determination of ^{210}Pb, ^{226}Ra and ^{137}Cs in beach sands along the coastline of Ghana. *African Journal of Environmental Pollution and Health, 9*, 17-23.

Papaefthymiou, H., Papatheodorou, G., Moustakli, A., Christodoulou, D., & Geraga, M. (2007). Natural radionuclides and ^{137}Cs distributions and their relationship with sedimentological processes in Patras Harbour, Greece. *Journal of Environmental Radioactivity, 94*(2), 55-74.

Pfitzner, J., Brunskill, G., & Zagorskis, I. (2004). ^{137}Cs and excess ^{210}Pb deposition patterns in estuarine and marine sediment in the central region of the Great Barrier Reef Lagoon, north-eastern Australia. *Journal of EnvironmentalRradioactivity, 76*(1), 81-102.

Qureshi, A. A., Tariq, S., Din, K. U., Manzoor, S., Calligaris, C., & Waheed, A. (2014). Evaluation of excessive lifetime cancer risk due to natural radioactivity in the rivers sediments of Northern Pakistan. *Journal of Radiation Research and Applied Sciences, 7*(4), 438-447.

Ravanat, J., Breton, J., Douki, T., Gasparutto, D., Grand, A., Rachidi, W., & Sauvaigo, S. (2014). Radiation-mediated formation of complex damage to DNA: a chemical aspect overview. *The British Journal of Radiology, 87*(1035), 20130715.

Sam, A., ElGanawi, A., Ahamed, M., & ElKhangi, F. (1998). Distribution of some natural and anthropogenic radionuclides in Sudanese harbour sediments. *Journal of Radioanalytical and Nuclear Chemistry, 237*(1-2), 103-107.

Santschi, P.H., Murray, J.W., Baskaran, M., Benitez-Nelson, C.R., Guo, L.D., Hung, C.C., Lamborg, C., Moran, S.B., Passow, U., & Roy-Barman, M. (2006). Thorium speciation in seawater. *Marine Chemistry*, *100*(3–4), 250-268.

Schmid, E., & Schrader, T. (2007). Different biological effectiveness of ionising and non-ionising radiations in mammalian cells. *Advances in Radio Science, 5*(1), 1-4.

Serigstad, B., Olsen, M., Mørk, T., Kristiansen, J., Wolinski, G., Gowing, S., Akam, D.N., & Whitehead, N. (2010). Marine environmental survey of bottom sediments in Cabinda and Soyo province, Angola. Cruise report No 4/2009.

Sirelkhatim, D., Sam, A., & Hassona, R. (2008). Distribution of ^{226}Ra–^{210}Pb–^{210}Po in marine biota and surface sediments of the Red Sea, Sudan. *Journal of Environmental Radioactivity, 99*(12), 1825-1828.

Sugandhi, S., Joshi, V. M., & Ravi, P. (2014). Studies on natural and anthropogenic radionuclides in sediment and biota of Mumbai Harbour Bay. *Journal of Radioanalytical and Nuclear Chemistry, 300*(1), 67-70.

SureshGandhi, M., Ravisankar, R., Rajalakshmi, A., Sivakumar, S., Chandrasekaran, A., & Anand, D.P. (2014). Measurements of natural gamma radiation in beach sediments of north east coast of Tamilnadu, India by gamma ray spectrometry with multivariate statistical approach. *Journal of Radiation Research and Applied Sciences, 7*(1), 7-17.

Ulanovsky, A., Pröhl, G., & Gómez-Ros, J. (2008). Methods for calculating dose conversion coefficients for terrestrial and aquatic biota. *Journal of Environmental Radioactivity, 99*(9), 1440-1448.

UNSCEAR. (2000). *Sources and effects of ionizing radiation: sources (Vol. 1). New York* (Vol. 1): United Nations Publications.

Xinwei, L., Lingqing, W., & Xiaodan, J. (2006). Radiometric analysis of Chinese commercial granites. *Journal of Radioanalytical and Nuclear Chemistry, 267*(3), 669-673.

Chapter 4

Assessment of DDT, HCH and PAH Contamination and Associated Ecotoxicological Risks in Surface Sediments of Coastal Tema Harbour (Ghana)

A modified version of this chapter was published as: Botwe, B.O., Kelderman, P., Nyarko, E., & Lens, P.N.L. (2017). Assessment of DDT, HCH and PAH contamination and associated ecotoxicological risks in surface sediments of coastal Tema Harbour (Ghana). *Marine Pollution Bulletin, 115, 480-488*.

Abstract

This study assessed DDTs, HCHs and PAHs contamination in sediments from the Tema Harbour (Ghana) and the associated ecotoxicological risks. The results showed widespread DDTs, HCHs and PAHs contamination in the harbour sediments with mean concentrations ranging from 6.0-12.8, 2.8-12.7 and 2,750-5,130 $\mu g.kg^{-1}$ d.w, respectively. The silt-clay and total organic carbon contents of the sediments poorly correlated with the pollutant concentrations. DDTs and HCHs contamination relate to past use of DDT and lindane, which under the anoxic harbour conditions resulted in disproportionately higher concentrations of *p,p'*-DDD and γ-HCH in the sediments. No conclusion could be drawn on the sources of PAHs as either petrogenic or pyrogenic. The pollutant concentrations in the harbour sediments, particularly γ-HCH, may pose high ecotoxicological risks. In comparison to a previous study, this study indicates there has been a considerable reduction in PAH contamination in the Tema Harbour since the last major oil spill in 2007.

4.1. Introduction

Chemical pollution of coastal marine ecosystems is a major global issue due to the adverse human health, as well as their ecological and socioeconomic implications, such as degradation of marine habitats, death of organisms and decline in biodiversity, contamination of food sources and deprivation of livelihoods (Islam and Tanaka, 2004; Mestres et al., 2010). Moreover, severely polluted ecosystems may be impossible or very costly to clean-up and restore. Many coastal marine ecosystems have been polluted (Islam and Tanaka, 2004), and many are under threat due the increasing coastal populations and developmental activities (Birch and Hutson, 2009). Chemical pollution of coastal marine ecosystems is often linked to intense human activities in the coastal zone, such as industrial production, shipping, crude oil extraction, agriculture and sewage disposal (Islam and Tanaka, 2004; Birch and Hutson, 2009; Smith et al., 2009; Mestres et al., 2010). Priority pollutants include dichlorodiphenyltrichloroethane (DDT), hexachlorocyclohexane (HCH) and polycyclic aromatic hydrocarbons (PAHs) due to the environmental persistence of these organic pollutants, their long-range atmospheric transport, bioaccumulation along the food chain, as well as their associated adverse biological effects such as toxicity, endocrine disruption, cancer, mutations and reproductive anomalies (Willett et al., 1998; Wang et al., 2001; Islam and Tanaka, 2004; King et al., 2004; De Luca et al., 2004; Nyarko et al., 2011).

DDT and HCH are synthetic organochlorine compounds (OCs) previously used widely in public health programmes and as agricultural pesticides. Although the use of DDT and HCH is currently banned or restricted in many parts of the world, their inherent environmental characteristics, illegal use and poor management of stockpiles of DDT and HCH result in their occurrence in the environment (Singh and Lal, 2009; Ntow and Botwe, 2011). Furthermore, DDT is still permitted for use in malaria control in malaria endemic tropical developing countries (Ntow and Botwe, 2011). Thus, continuous monitoring and assessment of these persistent organochlorine pollutants in the environment remain relevant.

The occurrence of PAHs in the environment typically relates to the release of crude oil and refined petroleum products (referred to as petrogenic sources) or combustion of organic matter and fossil fuel (referred to as pyrogenic sources) (Wang et al., 2001; Yunker et al., 2002; De Luca et al., 2004; Neff et al., 2005; Nyarko et al., 2011). PAHs which contain two or three fused benzene rings are classified as low molecular weight PAHs (LPAHs), while those having four to six fused benzene rings are classified as high molecular weight PAHs (HPAHs). Although PAHs commonly occur as a mixture of several hundreds of related compounds in the environment (Neff et al., 2005), the sixteen PAH compounds (PAH$_{16}$) designated as priority PAHs by the United States Environmental Protection Agency (USEPA) have become prime targets in many PAH contamination and bioremediation studies (Wang et al 2001; De Luca et al., 2004; King et al., 2004).

As the ultimate repository of chemical pollutants released into aquatic ecosystems (Burton, 2002; Birch and Hutson, 2009; Hu et al., 2009; Lin et al., 2009; Smith et al., 2009), sediments usually attain high levels of chemical pollutants and pose hazards to benthic organisms as well as other organisms in the aquatic food chain (Hong et al., 1995). Thus, there have been growing efforts to remediate contaminated sediments, commonly by dredging (Burton, 2002). Coastal marine harbours, particularly those associated with industrial and urban centres, are of major concern as such harbours serve as receptacles for a variety of pollutants released from industrial, urban, shipping, fishing and other anthropogenic activities (Simpson et al., 1996; Mestres et al., 2010; Lin et al., 2009; Smith et al., 2009). Furthermore, the low-energy hydrodynamics in harbours (Mestres et al., 2010) enhance the accumulation of sediment-associated pollutants. Thus, Lin et al. (2009) and Simpson et al. (1996) reported high levels of DDT and PAH, respectively, in harbour sediments. Dredging with subsequent disposal of contaminated harbour sediments also pose a threat to the receiving ecosystem as contaminated sediments are potential sources of pollutants to biota (Hong et al., 1995; Burton, 2002). Investigation of sediment contamination and associated ecological risks is, therefore, a priority for environmental regulators (Burton, 2002; Birch and Hutson, 2009). Prior to undertaking remedial actions, screening-level ecotoxicological risk assessments on sediments are conducted to identify: (i) pollutants of potential ecotoxicological concern, (ii) priority areas, and (iii) the potential to cause adverse ecological effects (Burton, 2002; Long et al., 2006; Birch and Hutson, 2009).

The coastal marine Tema Harbour, located along the Gulf of Guinea at Tema in Ghana, is an international harbour with considerable shipping and fishing activity. The construction of the Tema Harbour in 1960 coincided with the commencement of industrialisation in Ghana, specifically in Tema, which has since been the major industrial city of Ghana. Major industrial activities in Tema include alumina smelting, crude oil refining, paint, steel and cement production. Thus, the Tema Harbour has been subjected to industrial, shipping, fishing and other anthropogenic activities for nearly six decades. Notable incidents at the Tema Harbour during the past decade include two major oil spills with subsequent clean-up and a fire explosion at the shipyard, which destroyed oil pipeline installations of the Tema Oil Refinery (TOR) at the harbour (Ghana New Agency, 29 March 2005). The fire and oil spill incidents at the Tema Harbour, shipping and fishing activities, as well as discharges from the surrounding industries might have caused release of PAHs and other pollutants into the harbour. Thus, assessment of sediment contamination and associated ecotoxicological risk is indispensable, as follow-up of a preliminary study conducted in 2008 suggesting sediment PAH pollution in Tema Harbour (Gorleku et al., 2014). The aim of the present study was to assess the levels, spatial distributions and sources of DDTs, HCHs and PAH$_{16}$ in surface sediments from the Tema Harbour. In addition, the ecotoxicological risks were evaluated using biological effect-based numerical sediment quality guidelines and total toxicity equivalence approaches.

4.2.   Materials and methods

4.2.1. Sediment sampling

Surface sediment samples were collected with a stainless steel Ekman grab (3.5 L) from 21 locations in the Tema Harbour in January 2014. The monitoring stations were spread over the different compartments of the Tema Harbour: fourteen in the Main Harbour excluding rocky areas, two each in the Inner Fishing Harbour and the Outer Fishing Harbour; and three in the Canoe Basin (S15-S17) (Fig. 3.1). At each station, three grab samples were collected and the surficial sediments scooped with a stainless steel spoon from the upper 0-2 cm layer, which is considered representative of recently deposited sediments (Phillips, 2007). Each composite sediment sample was mixed and then split into two: one for analysis of grain size and the other for analyses of organic contaminants (DDTs, HCHs and PAHs), total organic carbon (TOC) and total nitrogen (TN). The sediment samples were finally taken into Rilsan® bags, stored in an ice-cooled box on the field and later transported to the Department of Marine and Fisheries Sciences laboratory of the University of Ghana, about 60 km away. In the laboratory, the sediment samples were stored in a freezer at -20 °C till analysis within 4 weeks. The redox potential (E_h) of the bottom water at each monitoring station was measured *in situ* by using a Hanna multi-parameter probe (HI 9829, Hanna Instruments, USA), taking measurements after the E_h values had stabilised (typically within 2 min).

4.2.2. Sediment analyses

The silt-clay (<63 μm) and sand (63-2000 μm) fractions in the sediment samples were determined on wet samples by the wet sieving method following Wang et al. (2001). Analyses of DDTs, HCHs, PAHs, TOC and TN were performed on freeze-dried and homogenised bulk sediment samples. TOC and TN contents were determined following the Walkley-Black wet oxidation (Schumacher, 2002) and the Kjeldahl (Sáez-Plaza et al., 2013) method, respectively, after decarbonisation with excess 1 M HCl. Analysis of DDTs, HCHs and PAHs was by Soxhlet-extraction following Wang et al. (2001) and Fung et al. (2005), with some modifications. Briefly, about 10 g of dry homogenised sediment samples were spiked with 100 μl of 1 μg/mL standard solutions of acenaphthylene-d_8, acenaphthene-d_{10}, fluorene-d_{10} (as internal standards for PAH) and PCB-204 and PCB-230 (as internal standards for DDTs and HCHs) before Soxhlet-extraction with 150 mL of dichloromethane/acetone (1:1 v/v) mixture for 16 h. Prior to the extraction, the sediment samples were mixed with 5 g of anhydrous granular sodium sulphate previously heated at 400 °C for 4 h and 2 g of activated copper powder to remove sulphur. Using a rotary evaporator, the extracts were then slowly evaporated to near dryness and the residue re-dissolved in 2 mL *n*-hexane for clean-up.

Clean-up of the extracts was carried out using glass chromatographic columns (20 x 1.0 cm) packed with 8 g of Davisil Grade 923 silica gel from Sigma Aldrich, pre-heated at 130 °C for

16 h. The chromatographic columns were first conditioned by eluting successively with 15 mL of dichloromethane and 20 mL of *n*-hexane, before the extracts were applied to the columns and eluted successively with 15 mL of *n*-hexane and 20 mL of *n*-hexane/dichloromethane (80:20) mixture. The eluates were collected, concentrated to near dryness by a pure stream of N_2 gas and finally reconstituted in 1 mL *n*-hexane for analysis of DDTs, HCHs and PAH_{16}. The PAH_{16} were analysed by using a Perkin Elmer AutoSystem gas chromatograph with flame ionisation detector (GC-FID, Agilent 6890 Series GC System), while DDTs and HCHs were analysed by using the same GC with a ^{63}Ni electron capture detector (GC-ECD). Identities of PAHs, DDTs and HCHs were confirmed using a GC with mass selective detector (GC-MSD).

4.2.3. Quality assurance and control

All sediment samples were analysed in triplicates. Chemicals, solvents and reagents used in the extraction were of HPLC grade. All containers used were either new or thoroughly cleaned. Prior to sample analysis, the GC instrument was calibrated with mixed standards for DDTs and HCHs (EPA608-Pest.Mix 1, Sigma-Aldrich), and PAH_{16} (16 Priority EPA PAHs, Campro Scientific GmbH, Germany). Routine analyses of procedural blanks, reagent blanks, spiked blanks, sample replicates and samples of a Standard Reference Material for sediments (SRM 1944) from the National Institute of Standards and Technology (NIST) were conducted to assess contamination and recovery of analytes. The target organic contaminants were not detected in both procedural and reagent blanks. Recoveries ranged between 81 and 98% for PAHs and between 78 and 94% for DDTs and HCHs, with relative standard deviations (RSD) below 5 % (see Table 4.1). Reported PAH, DDT and HCH concentrations were not corrected for recoveries.

4.2.4. Data treatment

Two-tailed Pearson's product-moment correlations were conducted to determine associations between the measured sediment parameters using the Statistical Package for Social Sciences (SPSS version 16.0) software. Descriptive statistics (mean and standard deviations) were performed in Windows Excel (2007) at the 95 % confidence interval. One-way analysis of variance (ANOVA) with multiple comparisons (Holm-Sidak and Kruskal-Wallis) tests were performed to assess significant differences in the spatial distribution of measured sediment

Table 4.1 Measured (n =3) and certified concentrations of target OC and PAH compounds in standard reference material (SRM 1944)

Compound	Measured value	Certified value	Recovery (%)	RSD (%)
OCs (μg.kg^{-1} d.w)				
p,p'-DDT	109 ± 4	119 ± 11	92	3.3
p,p'-DDE	81 ± 4	$86 \pm 12^*$	94	4.8
p,p'-DDD	84 ± 2	$108 \pm 16^*$	78	2.0
α-HCH	1.9 ± 0.1	$2.0 \pm 0.3^*$	94	4.3
PAHs (mg.kg^{-1} d.w)				
Naphthalene	1.34 ± 0.04	1.65 ± 0.31	81	3.0
Phenanthrene	5.11 ± 0.06	5.27 ± 0.22	97	1.1
Anthracene	1.58 ± 0.04	1.77 ± 0.33	89	2.3
Fluoranthene	8.72 ± 0.10	8.92 ± 0.32	98	1.1
Pyrene	9.44 ± 0.11	9.70 ± 0.42	97	1.2
Benz[a]anthracene	4.51 ± 0.07	4.72 ± 0.11	96	1.6
Chrysene	4.53 ± 0.10	4.86 ± 0.10	93	2.1
Benzo[b]fluoranthene	3.60 ± 0.04	3.87 ± 0.42	93	1.1
Benzo[k]fluoranthene	2.17 ± 0.04	2.30 ± 0.20	94	1.7
Benzo[a]pyrene	4.10 ± 0.05	4.30 ± 0.13	95	1.3
Benzo[ghi]perylene	2.68 ± 0.03	2.84 ± 0.10	94	1.1
Indeno[1,2,3-cd]pyrene	2.58 ± 0.04	2.78 ± 0.10	93	1.6
Dibenz[a,h]anthracene	0.39 ± 0.01	0.424 ± 0.069	92	2.6
Fluorene	0.80 ± 0.02	$0.85 \pm 0.03^*$	94	1.9
Acenaphthene	0.52 ± 0.03	$0.57 \pm 0.03^*$	92	4.8

RSD = relative standard deviation

*Provided as reference concentration instead of certified concentration parameters using SigmaPlot software (version 11.0). The distributions of ΣDDTs, ΣHCHs, ΣPAHs and mERMQ in the harbour were mapped using ArcGIS software (version 10.2.2).

4.3. Results

4.3.1. Sediment physicochemical characteristics

The measured physicochemical characteristics of the Tema Harbour sediments are presented in Table 4.2. The bottom water in the harbour was anoxic with E_h values ranging from -50 to -420 mV. The grain size composition of the harbour sediments revealed a predominance of the sandy (63-2000 μm) fraction with the mean silt-clay (<63 μm) fraction varying between 21 and 39 %. Sediments from the north-eastern corner (stations S1-S4, S13 and S14) and central parts of the Main Harbour had relatively higher silt-clay contents. The TOC content of the sediments ranged from 2.2 to 7.2 %, showing a significant variation ($p = 0.027$, ANOVA) in spatial distribution but without any clear trend. Stations with relatively high TOC content include S2, S7, S12, the Canoe Basin (S15-S17) and the Inner Fishing Harbour (S18 and S19). There was, however, no significant correlation ($r = -0.004$, $p > 0.05$) between the TOC and silt-clay contents. The mean sediment TN content varied from 0.8 to 1.4 % among the monitoring stations, but no significant variation ($p = 0.782$, ANOVA) was observed in the distribution of TN across the monitoring stations. There was also a poor correlation ($r = -0.18$, $p > 0.05$) between the silt-clay and TN contents. The mean TOC contents were 2-9 times higher than the mean TN contents, with a poor correlation ($r = -0.144$, $p > 0.05$) between the TOC and TN contents.

4.3.2. Levels and distribution of DDTs and HCHs in the Tema Harbour sediments

A widespread occurrence of DDTs (*p,p'*-DDT, *p,p'*-DDE and *p,p'*-DDD) and HCHs (α-HCH, β-HCH, γ-HCH and δ-HCH) was observed in the harbour sediments. The mean concentration ranges of *p,p'*-DDT, *p,p'*-DDE and *p,p'*-DDD were 1.6-3.5, 0.8-2.0 and 2.8-7.5 μg.kg^{-1} d.w, respectively (Table 4.2). Overall, the mean concentrations of ΣDDTs (i.e., *p,p'*-DDT + *p,p'*-DDE + *p,p'*-DDD) ranged from 6.0 to 12.8 μg.kg^{-1} d.w, with significant differences in the spatial distribution (one way ANOVA, $p < 0.001$): the north-easternmost part of the harbour being most contaminated (Fig. 4.1a). The ΣDDTs concentrations highly correlated with the concentrations of *p,p'*-DDT ($r = 0.77$, $p < 0.01$) and *p,p'*-DDD ($r = 0.96$, $p < 0.01$), and to a lesser extent, with *p,p'*-DDE ($r = 0.49$, $p < 0.05$). A significant correlation was observed between the *p,p'*-DDT and *p,p'*-DDD concentrations ($r = 0.62$, $p < 0.01$), but both *p,p'*-DDT (*r*

= 0.24, p > 0.05) and p,p'-DDD (r = 0.33, p > 0.05) exhibited poor correlations with p,p'-DDE. Both ΣDDTs (r = 0.58, p < 0.01) and p,p'-DDD (r = 0.54, p < 0.05) exhibited significant correlations with the silt-clay contents, but poor correlations with both TOC and TN contents in the sediments. Moreover, the concentrations of p,p'-DDT and p,p'-DDE exhibited poor correlations with the silt-clay, TOC and TN contents in the sediments.

Like DDTs, HCHs were ubiquitous in the harbour sediments. The mean concentration ranges of α-HCH, β-HCH, γ-HCH and δ-HCH were 0.6-2.6, 0.4-5.2, 1.1-5.8 and 0.2-1.8 μg.kg^{-1} d.w, respectively (Table 4.2). One way ANOVA revealed significant differences (p < 0.001) in the distribution of γ-HCH across the monitoring stations: relatively higher levels were found in the Inner Fishing Harbour, stations S15 (in the Canoe Basin) and S4 and S14 (in the Main Harbour). Overall, the mean concentrations of ΣHCHs (i.e., α-HCH + β-HCH + γ-HCH + δ-HCH) varied from 2.8 to 13.6 μg.kg^{-1} d.w. The spatial distribution pattern shows the north-eastern part of the harbour being most contaminated (Fig. 4.1b). Good correlations were found between the concentrations of ΣHCHs and α-HCH (r = 0.53, p < 0.05), β-HCH (r = 0.88, p < 0.01), γ-HCH (r = 0.86, p < 0.01) and δ-HCH (r = 0.60, p < 0.01). Among the HCH isomers, significant correlations were observed between α-HCH and γ-HCH (r = 0.43, p < 0.05), β-HCH and γ-HCH (r = 0.61, p < 0.01), and β-HCH and δ-HCH (r = 0.69, p < 0.01), while poor correlations were observed between α-HCH and β-HCH (r = 0.23, p > 0.05), α-HCH and δ-HCH (r = 0.09, p > 0.05), and γ-HCH and δ-HCH (r = 0.31, p > 0.05). No significant correlations were observed between the HCHs and concentrations of DDTs, the silt-clay content or the TOC and TN content.

4.3.3. Levels and distribution of PAHs in the Tema Harbour sediments

The USEPA 16 priority PAHs (PAH$_{16}$) namely naphthalene (Nap), acenaphthylene (Acy), acenaphthene (Ace), fluorene (Flu), phenanthrene (Phe), anthracene (Ant), fluoranthene (Fla), pyrene (Pyr), benzo[a]anthracene (BaA), chrysene (Chr), benzo[b]fluoranthene (BbF), benzo[k]fluoranthene (BkF); benzo[a]pyrene (BaP), benzo[g,h,i]perylene (BghiP), indeno[1,2,3-cd]pyrene (IP) and dibenz[a,h]anthracene (DahA) were frequently detected in the harbour sediment samples and their mean concentrations are presented in Table 4.3. Generally, phenanthrene recorded the highest mean levels (510-750 μg.kg^{-1} d.w; mean = 610 μg.kg^{-1} d.w), followed by acenaphthylene (410-660 μg.kg^{-1} d.w; mean = 500 μg.kg^{-1} d.w) and pyrene (330-750 μg.kg^{-1} d.w; mean = 456 μg.kg^{-1} d.w). Concentrations of summed PAHs

(ΣPAH_{16}) ranged from 2750-5130 $\mu g.kg^{-1}$ d.w with a mean of 3700 $\mu g.kg^{-1}$ d.w, with significant variations in the distribution of PAH_{16} across the monitoring stations (Kruskal-Wallis one way ANOVA on ranks, $p < 0.001$). The PAH distribution pattern shows that sediments in the Main Harbour are less contaminated by ΣPAH_{16} compared to those of the Fishing Harbour (see Fig. 4.1c).

4.4. Discussion

4.4.1. Physicochemical characteristics of the Tema Harbour sediments

Sediment attributes such as silt-clay and organic carbon contents influence organic pollutant uptake by sediments (Wang et al., 2001; Ahrens and Depree, 2004; De Luca et al., 2004; King et al., 2004; Wang et al., 2006; Lin et al., 2009). Furthermore, the redox conditions of sediments play a role in the biodegradation of organic matter (Caille et al., 2003) and thereby, the accumulation of organic matter in sediments. The results of the grain-size analysis showed that the Tema Harbour sediments were predominantly sandy, with the silt-clay fraction constituting 21-39 % (Table 4.2). This predominance of sand may be due to its large inputs by ocean water currents and movements of ships (Kelderman, 2012).

The TOC/TN ratios in the sediments, which ranged from 2.4 to 9.0 (Table 4.2), indicate a predominance of autochthonous organic matter in the harbour sediments, as organic matter of marine (autochthonous) sources is characterised by TOC/TN ratios ≤ 8.0 (Burdige, 2007; Tesi et al., 2007). This predominance of autochthonous organic matter may be linked to the fact that the harbour hardly receives direct riverine inflows, which could have introduced large quantities of terrestrial (allochthonous) organic matter into the harbour. There was lack of a good correlation between the silt-clay fractions and TOC content, possibly due to localised anthropogenic inputs of organic matter into the harbour, which will vary across the monitoring stations. Organic loading coupled with poor water circulation may have contributed to the prevailing anoxic condition in the harbour. Such anoxic conditions have the potential to retard the microbial decomposition of organic matter, as anaerobic biodegradation rates of organic matter (methane production or sulphate reduction) in marine sediments are usually slower than aerobic biodegradation (Caille et al., 2003). The anoxic condition at the inner station in the Canoe Basin (S15) is of concern as the E_h value was below the minimum established value of -300 mV for natural sediments (Ye et al., 2013).

4.4.2. Effect of silt-clay and TOC content on the DDTs, HCHs and PAHs distribution

This baseline study shows that DDTs (*p,p'*-DDT, *p,p'*-DDE and *p,p'*-DDD), HCHs (α-HCH, β-HCH, γ-HCH and δ-HCH) and the USEPA 16 priority PAHs (PAH$_{16}$) were ubiquitous in the Tema Harbour sediments. Different distribution patterns of ΣDDTs, ΣHCHs and ΣPAHs in sediments were observed across the monitoring sites: the most north-eastern part of the harbour is relatively more contaminated by ΣDDTs (Fig. 4.1a), ΣHCHs (Fig. 4.1b) and ΣPAHs (Fig. 4.1c). The strong affinity of fine grain sediments and organic carbon for organic pollutants (Hu et al., 2009; Lin et al., 2009) has often resulted in significant correlations (Maruya et al., 1996; Wang et al 2001; Yang et al., 2005) and consequently, the need to normalise organic pollutant concentrations using the TOC content is required for comparison (Birch and Hutson, 2009; Lin et al., 2009). In this study, only ΣDDTs, *p,p'*-DDD and naphthalene (Nap) exhibited moderate, but significant, correlations with the silt-clay contents of the sediments, while only benzo[k]fluoranthene (BkF) exhibited a moderate, but significant, correlation with TOC.

These poor correlations may be due to various reasons. According to Ahrens et al. (2004), the presumption that organic pollutants and/or TOC preferentially partition to the fine sediment fractions is not always holding and therefore, should not be generalised for all sediments. PAH, for example, has been shown to preferentially partition to the coarse grains of sediments (Wang et al., 2001). Moreover, in the Tema Harbour, different anthropogenic activities are carried out at different areas, which may largely influence the spatial distributions of the pollutants. Thus, stations S1 and S2 are close to the dry docks with vessel painting and sandblasting activities; station S4 (Oil Berth) is close to the point of discharge of crude oil and refined petroleum products; stations S10 and S11 are points of container operation in the harbour; fuel storage and fuelling of industrial fishing vessels is carried out in the Outer Fishing Harbour (S20 and S21); while the Canoe Basin (S15-S17) is dedicated to artisanal fishing operations. As mentioned earlier, localised anthropogenic inputs of organic matter into the harbour will have contributed to the observed poor correlations between sediment pollutants and TOC and/or silt-clay contents. For example, effluent discharges from nearby food canneries into the Main Harbour near the Shekete area (stations S7 and S12) could be important localised sources of organic matter in the Main Harbour. The Canoe Basin (S15-S17) and the Inner Fishing Harbour (S18 and S19) are also exposed to high loadings of organic waste from the operations of the fisher folk.

Table 4.2 Physicochemical parameters and mean DDTs and HCHs concentrations (μg.kg^{-1} d.w) in Tema Harbour sediments and SQGs

Monitoring station	Water depth (m)	Eh (mV)	Silt-clay (%)	TOC (%)	TN (%)	TOC/TN	p,p'-DDT	p,p'-DDE	p,p'-DDD	ΣDDTs	α-HCH	β-HCH	γ-HCH	δ-HCH	ΣHCHs
S1	9.5	-100	37.2	3.8	1.0	3.8	2.5	1.5	4.3	8.3	1.0	0.5	1.4	0.7	3.6
S2	7	-120	31.4	7.2	0.8	9.0	2.6	1.9	5.8	10.3	0.6	0.6	1.1	0.5	2.8
S3	9	-100	33.6	4.8	0.8	6.0	3.2	1.3	5.3	9.7	2.1	0.9	2.9	0.4	6.1
S4	8	-80	38.5	3.9	0.8	4.9	2.8	2.0	3.1	7.9	2.3	1.6	5.5	0.4	9.8
S5	8	-70	34.7	2.7	0.9	3.0	2.5	1.2	3.3	7.0	2.1	1.6	3.3	0.2	7.1
S6	8	-90	34.6	3.8	1.0	3.8	2.3	0.9	2.8	6.0	0.9	0.6	1.7	0.3	3.4
S7	8	-90	28.2	5.4	0.8	6.8	3.2	1.3	4.2	8.7	1.2	1.4	4.1	0.2	6.7
S8	8.5	-90	25.1	2.2	0.8	2.2	1.6	1.2	3.7	6.4	2.0	1.5	3.1	1.5	8.2
S9	9	-70	29.8	4.7	0.8	5.9	2.9	1.4	4.5	8.7	1.2	0.5	4.4	0.2	6.2
S10	10	-80	27.3	2.9	1.2	2.4	3.2	1.6	5.1	9.9	1.2	0.4	3.5	0.5	5.6
S11	9	-80	24.5	2.9	1.1	2.6	3.0	0.8	4.2	8.1	0.6	0.7	2.0	0.5	3.7
S12	9	-50	30.6	5.5	1.0	5.5	3.3	1.5	5.4	10.1	0.7	1.0	2.1	0.2	3.9
S13	8	-80	32.6	4.8	1.0	4.8	2.4	1.1	4.4	8.0	0.8	1.5	4.4	1.4	8.0
S14	10	-90	23.7	3.8	0.8	4.8	2.1	1.4	5.5	8.0	1.8	1.6	5.3	1.4	10.2
S15	3.5	-360	25.6	5.8	1.4	4.1	3.5	1.9	7.5	12.8	1.3	1.5	5.1	0.4	8.2
S16	4	-240	24.8	4.5	0.8	5.6	3.4	1.6	6.8	11.8	1.4	1.6	3.1	0.5	6.7
S17	5	-160	22.2	6.6	0.9	7.3	3.2	1.5	5.7	10.4	1.5	0.5	3.8	0.6	6.5
S18	8	120	28.8	6.7	1.0	6.7	2.7	1.2	4.9	8.8	1.1	5.2	5.8	1.4	13.6
S19	8	110	25.2	5.3	0.8	6.6	2.5	1.9	3.8	8.2	1.5	4.6	5.0	1.8	12.7
S20	8	-110	24.6	3.9	0.9	4.3	2.3	1.6	3.5	7.4	1.9	3.5	4.1	1.7	11.2
S21	8	-100	21.4	2.9	1.0	2.9	2.4	1.4	3.3	7.2	2.6	2.9	4.4	1.5	11.5
Mean	-	-	28.8	4.4	0.9	-	2.7	1.4	4.6	8.8	1.4	1.6	3.6	0.8	7.4
[a]ERL							1	2.2	2	1.58	-	-	0.32	-	-
[a]ERM							7	27	20	46.1	-	-	1	-	-

[a]Sediment Quality Guidelines (SQGs): ERL and ERM (μg.kg^{-1}) from Long et al. (2006); ERL = effective range low; ERM = effective range median.

Table 4.3 Mean concentrations ($\mu g.kg^{-1}$ d.w, $n = 3$) of USEPA 16 priority PAHs (PAH_{16}) in Tema Harbour surface sediments, SQGs and TEFs

Sample station	Low Molecular Weight PAHs (LPAHs)						High Molecular Weight PAHs (HPAHs)										$\Sigma LPAH_{16}$
	Nap	Acy	Ace	Fle	Phe	Ant	Fla	Pyr	BaA	Chr	BbF	BkF	BaP	BghiP	IP	DahA	
S1	270	420	270	150	530	26	310	340	60	270	130	64	160	50	35	13	3010
S2	380	410	240	170	550	28	230	350	100	360	150	53	100	14	38	19	3200
S3	350	420	260	160	570	35	350	350	100	430	100	84	140	50	36	27	3470
S4	390	590	350	350	750	37	440	650	170	520	160	64	160	120	150	26	4900
S5	330	490	240	150	550	26	240	350	100	320	140	52	160	17	50	24	3200
S6	360	410	230	150	530	33	320	340	70	300	130	69	160	16	19	33	3170
S7	260	410	150	140	530	29	230	340	60	220	120	61	150	14	15	18	2750
S8	250	420	250	160	530	30	210	330	100	310	170	46	140	40	27	30	3040
S9	340	450	260	170	550	27	230	340	70	270	140	56	110	40	50	21	3100
S10	290	440	230	140	590	33	240	400	100	300	130	44	130	20	12	43	3100
S11	260	480	240	140	540	28	230	350	100	280	160	82	150	32	32	29	3120
S12	260	470	250	150	560	30	230	350	100	290	150	55	110	13	40	22	3100
S13	250	500	150	140	510	24	230	340	60	240	110	43	150	26	28	23	2830
S14	350	470	250	150	560	24	330	350	90	370	130	43	140	28	15	42	3340
S15	490	660	380	190	750	46	440	750	170	580	130	43	170	200	130	28	5130
S16	440	630	330	350	740	38	430	730	150	550	110	51	140	120	130	34	4980
S17	460	620	340	340	730	41	420	700	160	530	120	40	160	190	120	20	5000
S18	360	500	250	140	560	30	330	340	80	380	110	67	150	22	22	34	3370
S19	370	580	290	360	750	32	430	660	160	510	130	53	150	130	140	21	4750
S20	360	550	270	360	730	40	430	600	160	500	150	49	170	110	130	29	4610
S21	330	600	170	340	750	32	440	620	160	470	130	40	150	120	140	18	4490
Mean	340	500	260	210	610	32	320	460	110	380	130	55	150	65	64	26	3700
ERL	160	44	16	19	240	85.3	600	665	261	384	-	-	430	-	-	63.4	4022
ERM	2100	640	500	540	1500	1100	5100	2600	1600	2800	-	-	1600	-	-	260	44792
TEF	0.001	0.001	0.001	0.001	0.001	0.01	0.001	0.001	0.1	0.01	0.1	0.1	1	0.01	0.1	5 (1*)	-

ERL and ERM values ($\mu g.kg^{-1}$) are from Burton (2002); TEF = toxic equivalency factor (Nisbet and LaGoy, 1992)

* USEPA value (https://fortress.wa.gov/ecy/clarc/FocusSheets/tef.pdf)

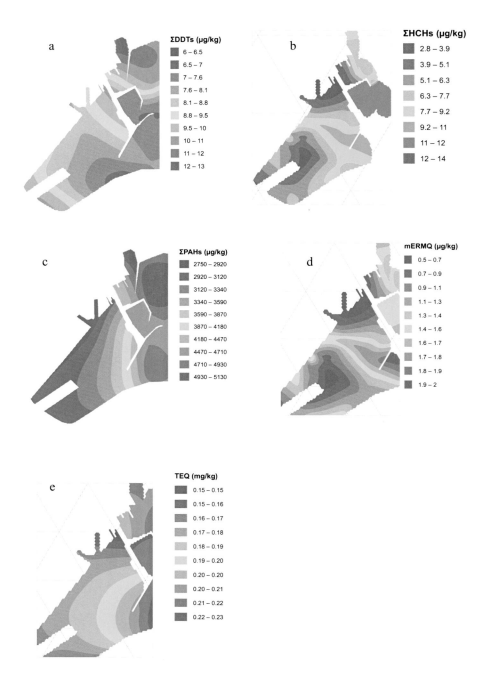

Fig. 4.1 Distribution patterns of (a) ΣDDTs, (b) ΣHCHs, (c) ΣPAHs, (d) mERMQ and (e) TEQ in Tema Harbour sediments

4.4.3. Sources of DDTs and HCHs in the Tema Harbour sediments

Both DDT and HCH are pesticides of anthropogenic origin: DDT is commercially produced as technical DDT and HCH as technical HCH and lindane. Although the Tema Harbour does not receive direct riverine inflows, it may be impacted by inland and coastal agricultural activities via alongshore water and sediment transport, as well as by atmospheric deposition and surface runoff (Botwe et al., 2012). As the Tema Harbour serves as the main sea port for imports, it is also possible that accidental pesticides spills may have occurred in the past, and have contributed to DDT and HCH contamination in the harbour sediments. Isolated cases of DDT use by Ghanaian fisher folks have also been reported (http://www.graphic.com.gh/news/general-news/danger-watch-the-fish-you-buy.html). There is, however, no information on how widespread this practice has been or on the quantities of DDT used. DDT is a key constituent in antifouling paints widely used in boat maintenance. This has earlier been identified as a major source of DDT contamination in fishing harbours (Lin et al., 2009) and coastal marine sediments (Hu et al., 2009; Yu et al., 2011). However, the use of these DDT-based antifouling paints on fishing vessels in Ghana has not been reported.

DDT and lindane are included in the list of persistent organic pollutants (POPs) targeted for global elimination under the Stockholm Convention, which Ghana has ratified and both pollutants were officially banned from use in Ghana in 1985 and 2002, respectively (Ntow and Botwe 2011). Prior to their ban in Ghana, they were largely used as pesticides in agriculture, e.g. lindane predominantly in cocoa plantations. Since their ban, residues of DDT and lindane have been measured in water, sediment, crops and biota in Ghana (Ntow and Botwe 2011), owing to their persistence, illegal use and existence of stockpiles (Singh and Lal 2009; Ntow and Botwe 2011). For example, in 2008, about 71,000 kg of banned pesticides were detected at a facility belonging to the Ghana Cocoa Board (Daily Graphic newspaper, 8 April 2008).

DDT and its degradation products DDE and DDD are all persistent in the environment with similar physical and chemical characteristics (WHO, 1989). Due to the low proportions of DDE and DDD in technical DDT, DDT/DDD and DDT/(DDE + DDD) ratios < 1.0 are generally indicative of past use of DDT, whereas ratios > 1.0 indicate current use (Li et al., 2014). The DDT/DDD and DDT/(DDE + DDD) ratios in all analysed Tema Harbour sediments were < 1.0 (Table 4.4), indicating that DDT contamination in the harbour is likely the result of DDT use in the distant past. A significant positive correlation ($p < 0.01$) was observed between p,p'-DDT and p,p'-DDD, which suggests some contribution of these components from influx of weathered sediment-bound DDT. Among the DDTs, disproportionately higher levels of p,p'-DDD (39-61 % of the measured ΣDDTs concentrations) were found in sediments at the monitoring stations (see Table 4.4), indicating substantial transformation of p,p'-DDT to p,p'-DDD in the harbour sediments. The higher levels of p,p'-DDD relative to those of p,p'-DDE can be attributed to the discriminatory degradation of p,p'-DDT to p,p'-DDD in anoxic sediments (Lin et al., 2009).

The α-HCH/γ-HCH ratio in sediments allows delineation of the sources of HCH as technical HCH (predominantly constituted by α-HCH) or lindane (>99 % γ-HCH). Technical HCH has α-HCH/γ-HCH ratios ranging between 3 and 7, whereas those of lindane are < 1.0 or close to 1.0 (Willett et al., 1998). The α-HCH/γ-HCH ratios in sediments from all the monitoring stations were < 1.0 (Table 4.4), pointing to the use of lindane as the major source of HCH contamination in the harbour. The γ-HCH isomer accounted for 37-71 % of the ΣHCHs concentrations in the sediments. In anaerobic sediments, as in those investigated, the conversion of γ-HCH to other HCH isomers can be an extremely slow process (Walker et al., 1999), which could have resulted in the predominance of γ-HCH in the anaerobic harbour sediments.

4.4.4. Sources of PAHs in the harbour sediments

Most of the PAHs exhibited significant correlations, possibly because they derive from common sources. Various anthropogenic activities within the harbour (e.g. crude oil spills during bunkering and fuelling of vessels, fuel leakages, ballast and bilge water discharges from ships, ship painting at the dry-dock, and vehicular emissions) as well as discharges and emissions from various industrial activities that characterise the city of Tema (e.g. oil refinery, aluminium smelting, cement production, pain manufacturing, and fish processing) may have contributed to the PAH contamination in the sediments (Yunker et al., 2000; Yunker et al., 2002; Neff et al., 2005). Discharges of oil into the harbour from the aforementioned sources may occur on a daily basis. Moreover, as mentioned, there have been two major incidents of oil spill in the Tema Harbour, in 2005 and 2007. A study conducted shortly after these incidents in 2008 reported mean ΣPAHs concentrations in the range of 28,600-190,000 µg.kg⁻

[1] w.w (Gorleku et al. 2014). Although differences in concentration units do not allow for comparison, we infer that the mean ΣPAHs concentrations in this study (2,750-5,130 µg.kg^{-1} d.w) conducted in 2014 are much lower than the levels reported by Gorleku et al. (2014). Thus, this study indicates that there has been a considerable reduction in PAH contamination in the Tema Harbour during the period 2008-2014.

To characterise the dominant sources of PAH contamination in the Tema Harbour sediments, ΣLPAH/ΣHPAH and three PAH isomeric ratios viz. anthracene/[anthracene + phenanthrene], fluoranthene/[fluoranthene + pyrene] and benzo(a)anthracene/[benzo(a)anthracene + chrysene] were evaluated (Table 4.4). Pyrogenic sources are mainly related to the combustion of organic matter and show enrichment in the high-molecular weight PAH (HPAH) components, whereas petrogenic sources relate to spillages of petroleum/crude oil and refined petroleum products and exhibit enrichment in the low-molecular weight PAH (LPAH) components (Yunker et al., 2000; Yunker et al., 2002; Neff et al., 2005). The ΣLPAH/ΣHPAH ratios in the sediments at the monitoring stations were close to 1.0, suggesting similar contributions of PAHs from both petrogenic and pyrogenic sources. The anthracene/[anthracene + phenanthrene] and fluoranthene/[fluoranthene + pyrene] ratios indicate PAH contamination in the harbour derived mainly from petrogenic sources, whereas the benzo(a)anthracene/[benzo(a)anthracene + chrysene] ratios suggest rather mixed petrogenic and pyrogenic sources. Thus, the use of different PAHs source diagnostic ratios has not yielded overall consistent results for the Tema Harbour sediments investigated.

Table 4.4 Source diagnostic DDT, HCH and PAH ratios in Tema Harbour sediments

Monitoring station	DDT/ DDE	DDT/ DDD	DDE/ DDD	DDT/ (DDE+DDD)	ΣLPAHs/ ΣHPAHs	Ant/ (Ant+Phe)	Fla/ (Fla+Pyr)	BaA/ (BaA+Chr)	α-HCH/ γ-HCH	β-HCH/ γ-HCH	δ-HCH/ γ-HCH
S1	1.6	0.6	0.4	0.4	1.2	0.05	0.5	0.18	0.7	0.4	0.5
S2	1.3	0.4	0.3	0.3	1.3	0.05	0.4	0.22	0.6	0.6	0.5
S3	2.5	0.6	0.2	0.5	1.1	0.06	0.5	0.19	0.8	0.3	0.1
S4	1.5	0.9	0.6	0.6	1.0	0.05	0.4	0.25	0.4	0.3	0.1
S5	2.1	0.8	0.4	0.6	1.2	0.05	0.4	0.24	0.6	0.5	0.1
S6	2.5	0.8	0.3	0.6	1.2	0.06	0.5	0.19	0.6	0.3	0.1
S7	2.4	0.8	0.3	0.6	1.2	0.05	0.4	0.21	0.3	0.3	0.05
S8	1.4	0.4	0.3	0.3	1.2	0.05	0.4	0.24	0.6	0.5	0.5
S9	2.1	0.6	0.3	0.5	1.4	0.05	0.4	0.21	0.3	0.1	0.5
S10	2.1	0.6	0.3	0.5	1.2	0.05	0.4	0.25	0.3	0.1	0.05
S11	3.6	0.7	0.2	0.6	1.2	0.05	0.4	0.26	0.2	0.4	0.1
S12	2.2	0.6	0.3	0.5	1.3	0.05	0.4	0.26	0.3	0.5	0.3
S13	2.2	0.6	0.3	0.4	1.3	0.04	0.4	0.20	0.2	0.3	0.1
S14	1.5	0.4	0.3	0.3	1.2	0.04	0.5	0.20	0.3	0.3	0.3
S15	1.9	0.5	0.3	0.4	1.0	0.06	0.4	0.23	0.3	0.3	0.3
S16	2.1	0.5	0.2	0.4	1.0	0.05	0.4	0.21	0.5	0.5	0.2
S17	2.2	0.6	0.3	0.5	1.0	0.05	0.4	0.23	0.4	0.1	0.2
S18	2.3	0.6	0.2	0.5	1.2	0.05	0.5	0.17	0.2	0.9	0.2
S19	1.3	0.7	0.5	0.4	1.0	0.04	0.4	0.24	0.3	0.9	0.4
S20	1.4	0.6	0.5	0.4	1.0	0.05	0.4	0.24	0.5	0.8	0.4
S21	1.7	0.7	0.4	0.5	1.0	0.04	0.4	0.25	0.6	0.7	0.3

4.4.5. Ecotoxicological risk characterisation

Two approaches were used to characterise the ecotoxicological risks posed by the pollutant levels in the harbour sediments, viz. the ones based on numerical sediment quality guidelines (SQGs) and the total toxicity equivalence (TEQ). In the former, the effect range low (ERL) and effect range median (ERM) were employed to characterise the potential ecotoxicological risk posed by individual pollutants in the sediments (Burton, 2002; Long et al., 2006). Both ERL and ERM have an appreciable level of predictability of sediment toxicity (Long et al., 1998), viz. low risk for pollutant concentrations at or below ERL, moderate risk for concentrations above ERL but below ERM, and high risk for concentrations above ERM (Birch & Hutson, 2009). At all the monitoring stations, the γ-HCH concentrations exceeded the ERM by over an order of magnitude (see Table 4.2), indicating that γ-HCH can potentially pose a high ecotoxicological risk and should, therefore, be of primary concern. Compared to this, levels of *p,p'*-DDE were below the ERL value (low ecotoxicological risk), while the levels of *p,p'*-DDT and *p,p'*-DDD were between their respective ERL and ERM values (moderate risks) (Table 4.2). The levels of naphthalene, acetylnaphthylene, acenaphthene, fluorene and phenanthrene were between their ERL and ERM values (Table 4.3) (moderate risks), whereas the levels of the remaining PAH$_{16}$ congeners were below their ERL values (low risks). Since there are no ERLs and ERMs for α-HCH, β-HCH, δ-HCH, benzo[b]fluoranthene, benzo[k]fluoranthene, benzo[g,h,i]perylene and indeno[1,2,3-cd]pyrene, the potential ecotoxicological risk posed by these pollutants cannot be characterised using this approach.

Considering that different pollutants in sediments may potentially act in concert to cause toxicity different from that of the individual pollutants (Burton, 2002; Long et al., 2006), the mean ERM quotient (mERMQ) approach (Long et al., 2006; Birch & Hutson, 2009) was further adopted to characterise the integrated ecotoxicological risks due to the combined effect of PAHs, DDTs and γ-HCH mixtures in the Tema Harbour sediments investigated. The mERMQs were evaluated for each sediment sample by normalising the levels of ΣPAHs, ΣDDTs and γ-HCH to their respective ERM values, then summing the quotients and dividing by 3 (Long et al., 2006; Birch and Hutson, 2009). The assumptions here are that (1) PAHs, DDTs and γ-HCH contribute additively to the overall toxicity, rather than antagonistically or synergistically, and (2) samples with the same mERMQ pose similar ecotoxicological risks (Long et al., 2006; Birch and Hutson, 2009). Further adopting the risk classification by Birch and Hutson (2009), i.e. minimal risk if mERMQ < 0.1, low risk if $0.1 \leq$ mERMQ < 0.5, moderate risk if $0.5 \leq$ mERMQ < 1.5, and high risk if mERMQ $\geq$ 1.5. The results showed that the Tema Harbour sediments can potentially pose moderate to high ecotoxicological risks as sediments from 52% of the monitoring stations had mERMQ values exceeding 1.5, while those from the remaining stations had mERMQ values in the 0.5-1.5 range (Table 4.5). The spatial pattern of mERMQ (Fig. 4.1d) shows that large areas of the Fishing Harbour and the Main Harbour are relatively higher risk areas which deserve further management attention.

To provide an integral characterisation for PAH contamination of the harbour sediments, the total toxicity equivalence (TEQ) approach was used. TEQs at the different stations were calculated according to Eq. 1 (Nisbet and LaGoy, 1992):

$$TEQ \ (mg.kg^{-1}) = \Sigma(C_n \ x \ TEF_n) \tag{1}$$

where C_n and TEF_n are, respectively, the concentration $(mg.kg^{-1})$ and toxic equivalency factor (TEF) of an individual PAH congener in the sediment sample. TEFs of the PAH congeners express their toxicity relative to that of benzo(a)pyrene (Nisbet and LaGoy, 1992) and are presented in Table 4.3. Sediments can then be classified as uncontaminated for TEQ < 0.1 $mg.kg^{-1}$, slightly contaminated for 0.1 < TEQ < 1 $mg.kg^{-1}$, and significantly contaminated for TEQ $\geq$ 1.0 $mg.kg^{-1}$ (Yang et al., 2014). Table 4.5 shows that the TEQs associated with Tema Harbour sediments ranged from 0.2-0.4 $mg.kg^{-1}$ (mean value 0.3 $mg.kg^{-1}$). All TEQ values were within the 0.1-1.0 $mg.kg^{-1}$ range, indicating that the harbour sediments are moderately contaminated by the PAH mixtures (Yang et al., 2014).

The TEQ approach by the USA Environmental Protection Agency (USEPA) for the assessment of soil contamination by PAH mixtures under Method B of Model Toxics Control Act (MTCA) Cleanup Regulation (https://fortress.wa.gov/ecy/clarc/FocusSheets/tef.pdf) was further adopted. This approach is similar to the one by Nisbet and LaGoy (1992), see Eq. 1, except that a TEF value of 1.0, instead of 5.0, is assigned to dibenz[a,h]anthracene (Table 4.3). Following the USEPA approach, sediments with TEQ values $\geq$ 0.137 $mg.kg^{-1}$ are considered potentially toxic and harmful to environmental health, and therefore require clean-up. The results show that the TEQ values for all the harbour sediments exceeded 0.137 $mg.kg^{-1}$, ranging from 0.164-0.255 $mg.kg^{-1}$ (Table 4.5). This indicates that PAH contamination in the harbour sediments can potentially pose unacceptable risks to ecosystem health. Sediments in

Table 4.5 Calculated total toxicity equivalence (TEQ) and mean ERM quotient (mERMQ) values for Tema Harbour sediments

Monitoring station	[a]TEQ (mg.kg^{-1})	[b]TEQ (mg.kg^{-1})	mERMQ
S1	0.3	0.164	0.6
S2	0.2	0.172	0.5
S3	0.3	0.176	1.1
S4	0.4	0.195	1.9
S5	0.3	0.199	1.2
S6	0.4	0.203	0.6
S7	0.3	0.205	1.5
S8	0.3	0.206	1.1
S9	0.3	0.210	1.6
S10	0.4	0.217	1.3
S11	0.3	0.219	0.7
S12	0.3	0.221	0.8
S13	0.3	0.222	1.5
S14	0.4	0.224	1.9
S15	0.4	0.225	1.8
S16	0.4	0.228	1.2
S17	0.3	0.231	1.4
S18	0.4	0.238	2.0
S19	0.3	0.248	1.8
S20	0.4	0.255	1.5
S21	0.3	0.255	1.6
Mean	0.3	0.215	1.3
Threshold	0.1	0.137	0.1

[a]Nisbet and LaGoy (1992) method

[b]USEPA method (https://fortress.wa.gov/ecy/clarc/FocusSheets/tef.pdf)

the Main Harbour may be associated with relatively lower risks compared to the other parts of the Tema Harbour (Fig. 4.1e).

It should be noted that the ecotoxicological risk assessment was at a screening-level and only established the likelihood of DDTs, HCHs and PAH_{16} contamination in the harbour sediments to cause undesirable biological effects. Information on the hazard associated with harbour sediments from e.g. whole sediment bioassays, which integrate the effects of other pollutants such as heavy metals, are required to inform decisions about the handling of the harbour sediments. This baseline study nevertheless clearly shows there is a need to step up pollution control, prevention and management efforts by the Tema Harbour and responsible government authorities.

4.5. Conclusion

In conclusion, DDTs (p,p'-DDT, p,p'-DDE and p,p'-DDD), HCHs (α-HCH, β-HCH, γ-HCH and δ-HCH) and the USEPA 16 priority PAHs (PAH_{16}) are ubiquitous in the Tema Harbour sediments. The harbour sediments were predominantly sandy, with no significant correlations between the silt-clay/TOC contents and the concentrations of most of the pollutants investigated, probably due to the temporal and spatial variable contaminant loadings as well as redistribution by water currents. DDT in the Tema Harbour sediments has undergone considerable weathering to p,p'-DDD and thus, DDT contamination in sediments is attributable to the impact of past use of DDT. The concentrations of γ-HCH in Tema Harbour sediments were higher than those of α-HCH, β-HCH and δ-HCH, suggesting that HCH contamination relates to past use of lindane. The PAH_{16} concentrations in the sediments may pose appreciable ecotoxicological risks. γ-HCH was identified as a contaminant of potential concern. This baseline study has shown that there is a need to regulate discharges of PAHs into the harbour, considering their high potential toxicity.

References

Ahrens, M.J. & Depree, C.V., 2004. Inhomogeneous distribution of polycyclic aromatic hydrocarbons in different size and density fractions of contaminated sediment from Auckland Harbour, New Zealand: an opportunity for mitigation. *Mar. Pollut. Bull.* *48*(3), 341-350.

Botwe, B. O., Nyarko, E., & Ntow, W. J. (2012). Pesticide Contamination in Groundwater and Streams Draining Vegetable Plantations in the Ofinso District, Ghana. In: Hernandez-Soriano, M.C. (Ed.), Soil Health and Land Use Management, InTech, pp. 51-66. Rijeka, Croatia.

Birch, G.F. & Hutson, P., 2009. Use of sediment risk and ecological/conservation value for strategic management of estuarine environments: Sydney estuary, Australia. *Environ. Manage. 44*(4), 836-850.

Burdige, D.J., 2007. Preservation of organic matter in marine sediments: controls, mechanisms, and an imbalance in sediment organic carbon budgets? *Chem. Rev. 107*(2), 467-485.

Burton, G.A. Jr., 2002. Sediment quality criteria in use around the world. *Limnology, 3*(2), 65-76.

Caille, N., Tiffreau, C., Leyval, C. & Morel, J.L., 2003. Solubility of metals in an anoxic sediment during prolonged aeration. *Sci. Total Environ. 301*(1), 239-250.

De Luca, G., Furesi, A., Leardi, R., Micera, G., Panzanelli, A., Piu, P.C. & Sanna, G., 2004. Polycyclic aromatic hydrocarbons assessment in the sediments of the Porto Torres Harbour (Northern Sardinia, Italy). *Mar. Chem. 86*(1), 15-32.

Fung, C., Zheng, G., Connell, D., Zhang, X., Wong, H., Giesy, J.P., Fang, Z. & Lam, P.K.S., 2005. Risks posed by trace organic contaminants in coastal sediments in the Pearl River Delta, China. *Mar. Pollut. Bull. 50*(10), 1036-1049.

Gorleku, M.A., Carboo, D., Palm, L.M.N., Quasie, W.J. & Armah, A.K., 2014. Polycyclic aromatic hydrocarbons (PAHs) pollution in marine waters and sediments at the Tema Harbour, Ghana. *Acad. J. Environ. Sci.* **2**(7), 108-115.

Hong, H., Xu, L.-J., Zhang, L., Chen, J., Wong, Y. & Wan, T., 1995. Special guest paper: environmental fate and chemistry of organic pollutants in the sediment of Xiamen and Victoria Harbours. *Mar. Pollut. Bull. 31*(4), 229-236.

Hu, W., Wang, T., Khim, J. S., Luo, W., Jiao, W., Lu, Y., Naile, J.E., Chen, C., Zhang, X. & Giesy, J.P., 2010. HCH and DDT in sediments from marine and adjacent riverine areas of North Bohai Sea, China. *Arch. Environ. Contam. Toxicol. 59*(1), 71-79.

Islam, M.S. & Tanaka, M., 2004. Impacts of pollution on coastal and marine ecosystems including coastal and marine fisheries and approach for management: a review and synthesis. *Mar. Pollut. Bull. 48*(7-8), 624-649.

Kelderman, P., 2012. Sediment pollution, transport, and abatement measures in the city canals of Delft, The Netherlands. *Water Air Soil Pollut. 223*(7), 4627-4645.

King, A., Readman, J. & Zhou, J., 2004. Dynamic behaviour of polycyclic aromatic hydrocarbons in Brighton marina, UK. *Mar. Pollut. Bull. 48*(3), 229-239.

Li, P., Wang, Y., Huang, W., Yao, H., Xue, B. & Xu, Y., 2014. Sixty-year sedimentary record of DDTs, HCHs, CHLs and endosulfan from emerging development gulfs: a case study in the Beibu Gulf, South China Sea. *Bull. Environ. Contam. Toxicol. 92*(1), 23-29.

Lin, T., Hu, Z., Zhang, G., Li, X., Xu, W., Tang, J. & Li, J., 2009. Levels and mass burden of DDTs in sediments from fishing harbours: the importance of DDT-containing antifouling paint to the coastal environment of China. *Environ. Sci. Technol. 43*(21), 8033-8038.

Long, E.R., Field, L.J. & MacDonald, D.D., 1998. Predicting toxicity in marine sediments with numerical sediment quality guidelines. *Environ Toxicol and Chem. 17*(4), 714-727.

Long, E.R., Ingersoll, C.G. & MacDonald, D.D., 2006. Calculation and uses of mean sediment quality guideline quotients: a critical review. *Environ. Sci. Technol. 40*(6), 1726-1736.

Maruya, K.A., Risebrough, R.W. & Horne, A.J., 1996. Partitioning of polynuclear aromatic hydrocarbons between sediments from San Francisco Bay and their porewaters. *Environ. Sci. Technol. 30*(10), 2942-2947.

Mestres, M., Sierra, J., Mösso, C. & Sánchez-Arcilla, A., 2010. Sources of contamination and modelled pollutant trajectories in a Mediterranean harbour (Tarragona, Spain). *Mar. Pollut. Bull. 60*(6), 898-907.

Neff, J.M., Stout, S.A. & Gunster, D.G., 2005. Ecological risk assessment of polycyclic aromatic hydrocarbons in sediments: identifying sources and ecological hazard. *Integr. Environ. Assess. Manage. 1*(1), 22-33.

Nisbet, I.C. & LaGoy, P.K., 1992. Toxic equivalency factors (TEFs) for polycyclic aromatic hydrocarbons (PAHs). *Regul. Toxicol. Pharm. 16*(3), 290-300.

Ntow, W.J & Botwe, B.O., 2011. Contamination status of organochlorine pesticides in Ghana. In: Loganathan, B.G. & Lam, P.K.S. (Eds), Global Contamination Trends of Persistent Organic Chemicals. pp. 393-411. CRC Press, USA.

Nyarko, E., Botwe, B.O. & Klubi, E., 2011. Polycyclic aromatic hydrocarbons (PAHs) levels in two commercially important fish species from the coastal waters of Ghana and their carcinogenic health risks. *West Afr. J. Appl. Ecol. 19*(1), 53-66.

Phillips, C.R., 2007. Sediment contaminant patterns within coastal areas of the Southern California Bight: multivariate analyses of Bight'98 Regional Monitoring Data. *Bull. S. California Acad. Sci. 106*(3), 163-178.

Sáez-Plaza, P., Michałowski, T., Navas, M.J., Asuero, A.G. & Wybraniec, S., 2013. An overview of the Kjeldahl method of nitrogen determination. Part I. Early history, chemistry of the procedure, and titrimetric finish. *Crit. Rev. Anal. Chem. 43*(4), 178-223.

Schumacher, B.A., 2002. Methods for the determination of total organic carbon (TOC) in soils and sediments. USEPA Environmental Sciences Division National Exposure Research Laboratory, Ecological Risk Assessment Support Centre, Office of Research and Development. pp. 1-23. Las Vegas, USA

Simpson, C.D., Mosi, A.A., Cullen, W.R. & Reimer, K.J., 1996. Composition and distribution of polycyclic aromatic hydrocarbon contamination in surficial marine sediments from Kitimat Harbour, Canada. *Sci. Total Environ. 181*(3), 265-278.

Singh, A., & Lal, R. (2009). *Sphingobium ummariense* sp. nov., a hexachlorocyclohexane (HCH)-degrading bacterium, isolated from HCH-contaminated soil. *Int. J. Syst. Evol. Microbiol. 59*(1), 162-166.

Smith, J., Lee, K., Gobeil, C. & Macdonald, R., 2009. Natural rates of sediment containment of PAH, PCB and metal inventories in Sydney Harbour, Nova Scotia. *Sci. Total Environ. 407*(17), 4858-4869.

Tesi, T., Miserocchi, S., Goñi, M.A., Langone, L., Boldrin, A. & Turchetto, M., 2007. Organic matter origin and distribution in suspended particulate materials and surficial sediments from the western Adriatic Sea (Italy). *Estuar. Coast. Shelf Sci. 73*(3), 431-446.

Walker, K., Vallero, D.A. & Lewis, R.G., 1999. Factors influencing the distribution of lindane and other hexachlorocyclohexanes in the environment. *Environ. Sci. Technol. 33*(24), 4373-4378.

Wang, X.-C., Sun, S., Ma, H.-Q. & Liu, Y., 2006. Sources and distribution of aliphatic and polyaromatic hydrocarbons in sediments of Jiaozhou Bay, Qingdao, China. *Mar. Pollut. Bull. 52*(2), 129-138.

Wang, X.-C., Zhang, Y.-X. & Chen, R.F., 2001. Distribution and partitioning of polycyclic aromatic hydrocarbons (PAHs) in different size fractions in sediments from Boston Harbour, United States. *Mar. Pollut. Bull. 42*(11), 1139-1149.

WHO (World Health Organisation), 1989. DDT and its derivatives: Environmental aspects. Environmental Health Criteria 83. Geneva.

Willett, K.L., Ulrich, E.M. & Hites, R.A., 1998. Differential toxicity and environmental fates of hexachlorocyclohexane isomers. *Environ. Sci. Technol. 32*(15), 2197-2207.

Yang, R.-q., Jiang, G.-b., Zhou, Q.-f., Yuan, C.-g. & Shi, J.-b., 2005. Occurrence and distribution of organochlorine pesticides (HCH and DDT) in sediments collected from East China Sea. *Environ. Int. 31*(6), 799-804.

Yang, Y., Woodward, L.A., Li, Q. X. & Wang, J., 2014. Concentrations, Source and Risk Assessment of Polycyclic Aromatic Hydrocarbons in Soils from Midway Atoll, North Pacific Ocean. *PLoS One, 9*(1): 1-7.

Ye, S., Laws, E.A. & Gambrell, R., 2013. Trace element remobilisation following the resuspension of sediments under controlled redox conditions: City Park Lake, Baton Rouge, LA. *Appl. Geochem. 28*, 91-99.

Yu, H.-Y., Shen, R.-L., Liang, Y., Cheng, H. & Zeng, E.Y., 2011. Inputs of antifouling paint-derived dichlorodiphenyltrichloroethanes (DDTs) to a typical mariculture zone (South China): potential impact on aquafarming environment. *Environ. Pollut. 159*(12), 3700-3705.

Yunker, M.B., Macdonald, R.W., Brewer, R., Sylvestre, S., Tuominen, T., Sekela, M., Mitchell, R.H., Paton, D.W., Fowler, B.R., Gray, C., Goyette, D. & Sullivan, D., 2000. Assessment of natural and anthropogenic hydrocarbon inputs using PAHs as tracers. The Fraser River basin and Strait of Georgia, 1987-1997. Burrard Inlet Environmental Action Programme, Canada.

Yunker, M.B., Macdonald, R.W., Vingarzan, R., Mitchell, R.H., Goyette, D. & Sylvestre, S., 2002. PAHs in the Fraser River basin: a critical appraisal of PAH ratios as indicators of PAH source and composition. *Org. Geochem. 33*(4), 489-515.

Chapter 5

Metal Distribution and Fractionation in Surface Sediments of Coastal Tema Harbour (Ghana) and its Ecological Implications

A modified version of this chapter was published as: Botwe, B.O., Alfonso, L., Nyarko, E., & Lens, P.N.L. (2017). Metal Distribution and Fractionation in Surface Sediments of Coastal Tema Harbour (Ghana) and its Ecological Implications. *Environmental Earth Sciences, 76, 514-530.*

Abstract

This study investigated the distribution and fractionation of metals (Mn, Ni, Pb, Cr, Cu, Zn, As, Cd, Hg and Sn) in surface sediments of Tema Harbour (Greater Accra, Ghana) as well as its ecological implications. Significant differences in sediment concentrations of Mn, Ni, Cr, Cu, Zn, As and Sn were observed across the Tema Harbour. Geochemical indices indicate that Cd, Hg, Pb, Cu, Zn, As and Sn in the Tema Harbour sediments derived mainly from anthropogenic sources, while Mn, Ni and Cr were mainly of lithogenic origin. Metal fractionation revealed a predominance of Al, Mn, Ni, Pb, Cr, Cu, As and Sn in the residual phase. In contrast, Cd and Hg were mainly present in the exchangeable phase, while Zn was mainly associated with the reducible phase. Based on the metal fractionation in the Tema Harbour sediments, the potential risks of metal bioavailability were high for Cd and Hg, low-medium for Mn, Ni, Zn, As and Sn, and low for Pb, Cr, and Cu. A screening-level ecotoxicological assessment revealed high potential toxicity of Hg and moderate potential toxicities of Pb, Cu, Zn, As and Cd in the Tema Harbour sediments. The potential influence of the buffer intensity, silt-clay, total organic carbon and carbonate content on the metal distribution in the Tema Harbour sediments were also inferred from their correlations. Comparison with previous studies did not reveal a progressive increase in metal contamination at the Tema Harbour since the year 2000.

5.1. Introduction

Harbour sediments are often contaminated with a wide range of pollutants including metals (Casado-Martinez et al. 2006; Palma and Mecozzi 2007), which derive from anthropogenic activities associated with harbours such as shipping, fishing as well as industrial and urban wastewater discharges (Lepland et al. 2010; Schipper et al. 2010; Nyarko et al. 2014; Botwe et al. 2017a). Polluted sediments are potential secondary sources of chemical pollutants to benthic organisms living in contact with sediments and ultimately to other organisms within the food chain (Burton 2002; Kelderman and Osman 2007; Nyarko et al. 2011; Botwe et al. 2017a). Metals such as Ni, Pb, Cr, Cu, Zn, As, Cd and Hg are hazardous to marine organisms (Casado-Martinez et al. 2006; Schipper et al. 2010) and their contamination in the environment has adverse human health, socioeconomic and food security implications (Birch and Hutson 2009; Lepland et al. 2010; Botwe et al. 2017a). Contaminated harbour sediments may adversely impact other coastal marine and terrestrial ecosystems via the disposal of dredged materials (Caille et al. 2003; Schipper et al. 2010; Ho et al. 2012a; Botwe et al. 2017a).

Considering the threat contaminated sediments may pose to aquatic and terrestrial ecosystems, remediation of contaminated sediments is a priority for environmental management in harbours (Pozza et al. 2004). Remediation of harbour sediments requires a prior screening of the sediments for their potential ecotoxicity and environmental risks in order to obtain information on (1) pollutants of prime concern, (2) the potential to cause adverse biological effects in benthic organisms and (3) identify the most impacted areas that deserve priority attention (Burton 2002; Long et al. 2006; Birch and Hutson 2009; Botwe et al. 2017a). A screening-level ecotoxicological assessment of sediments involves comparisons of measured pollutant concentrations in the sediment with established biological effect-based sediment quality guidelines (SQGs) (Long et al. 1995; Burton 2002; Botwe et al. 2017a). Two SQGs which have been widely used around the world for ecotoxicological screening of sediments are the effects-range low (ERL) and effects-range median (ERM) SQGs (Long et al. 1995; Burton 2002; Botwe et al. 2017a).

Sediment-bound metals may associate with different geochemical phases present in sediments, e.g. carbonate, Fe-Mn oxides and hydroxides, organic matter and sulphides. The phase association (fractionation) of metals in sediments determines their binding strength and consequently, their potential mobility and bioavailability (Calmano et al. 1993; Pueyo et al. 2001; Kelderman and Osman 2007; Ho et al. 2012b; Hamzeh et al. 2014). Investigation of the metal distribution over different geochemical phases (metal fractionation) in sediments is, therefore, relevant when screening the sediments for their potential environmental risks. A standardised procedure widely used for metal fractionation in sediments is the Community Bureau of Reference (BCR) three-step sequential extraction technique, which yields four metal fractions, namely the acid-soluble (exchangeable and bound to carbonate), reducible (bound to Fe-Mn oxides and hydroxides), oxidisable (bound to organic matter and sulphides) and residual (bound within the crystal lattice of minerals) fractions (Quevauviller et al. 1997; Pueyo et al. 2001; Ho et al. 2012a). Since metal fractionation and potential mobility in sediments are influenced by environmental factors such as pH, buffer intensity and redox potential (Calmano et al. 1993; Kelderman and Osman 2007; Palma and Mecozzi 2007; Ho et al. 2012a; Hamzeh et al. 2014), these factors should be considered in characterising the potential environmental risk of metal-contaminated sediments.

The coastal marine Tema Harbour in Greater Accra (Ghana) was constructed in 1960 and commissioned in 1962 to facilitate international trade and boost industrial development in Ghana. Thus, the Tema Harbour was sited at the seacoast of the industrial city of Tema, Greater Accra in Ghana and its commissioning marked the beginning of Ghana's industrial era. With over 180 industries in Tema, as well as shipping and fishing activities at the Tema Harbour, the Tema Harbour is susceptible to metal pollution (Lepland et al. 2010; Schipper et al. 2010; Nyarko et al. 2014). Previous studies in 2000 (http://open_jicareport.jica.go.jp/pdf/11681632_03.pdf) and 2011 (Nyarko et al. 2014) characterised metal contamination levels in the Tema Harbour. However, the above studies focused on total metal concentrations without examining metal fractionation in the Tema Harbour sediments. Moreover, the Tema Harbour sediments have not been characterised for their buffer intensity against acidification, which determines metal leachability and associated environmental risk (Calmano et al. 1993; Jain 2004; Iqbal et al. 2013; Ho et al. 2012a). The objective of this study was, therefore, to investigate the distribution and fractionation of metals (Mn, Ni, Pb, Cr, Cu, Zn, As, Cd, Hg and Sn) in surface sediments from the Tema Harbour as well as its ecological implications.

5.2. Materials and methods

5.2.1. Sediment sampling and sample analyses

Surface sediment samples were collected in November 2014 with a stainless steel Ekman grab (3.5 L) from 21 stations across the different compartments of the Tema Harbour, which were previously monitored for organic (Botwe et al. 2017a) and radioactivity (Botwe et al. 2017b) contamination. Thus, sediment samples were collected from fourteen stations in the Main Harbour, two stations each in the Inner Fishing Harbour and the Outer Fishing Harbour, and three stations in the Canoe Basin (see Fig. 3.1). The Main Harbour is where ships call at the Tema Harbour, and has a total of 14 berths. The Inner Fishing Harbour and the Outer Fishing Harbour provide handling facilities for semi-industrial and industrial fishing vessels, while the Canoe Basin serves as an artisanal canoe fishing landing site. Adopting the sampling strategy by Botwe et al. (2017a), three grab samples were collected at each station. Only grabs that arrived on the deck firmly closed without water leakage and were not filled with sediment to the lid were sampled to ensure that the fine particles were not lost and the surface layer was intact (Aloupi and Angelidis 2001; Botwe et al. 2017b).

From each grab, two sub-samples each of about 200 g wet weight were taken with acid-washed plastic spoons from the upper 0-2 cm layer, which is representative of recently deposited sediments (Phillips 2007; Botwe et al. 2017a), into acid-washed FoodSaver® zipper bags. One subsample was for the analyses of metals, total organic carbon (TOC), carbonate, pH (and subsequently buffer intensity) and the other for grain size analysis. Samples for metal analysis were vacuum-sealed to minimise oxidation. To avoid contamination, sediments in direct contact with the grab were not sampled (Botwe et al. 2017b). Redox potential was measured *in situ* near the sediment-water interface using a Hanna multiparameter probe (HI 9829) following Botwe et al. (2017b). All sediment samples were kept on ice and transported to the University of Ghana laboratory for further analysis.

In the laboratory, the silt-clay (<63 µm) and sand (63-2000 µm) fractions in the sediment samples were determined on wet samples by the wet sieving method following the method by Wang et al. (2001). Sediment samples for metal, total organic carbon (TOC), carbonate and pH analyses were freeze-dried and then homogenised using a Fritsch Pulveriser. Total metal concentrations and metal fractionation were determined on about 0.2 g portions of freeze-dried homogenised sediment samples by adopting the harmonised Community Bureau of Reference (BCR) 3-step sequential extraction and *aqua regia* extraction techniques, respectively, following Pueyo et al. (2001). Metal concentrations were assayed using an Atomic Absorption Spectrometer (Varian AA 240FS). The detection limits for the measured metals were: 0.10 for Al, Hg and As; 0.20 for Cd; 0.50 for Mn, Ni, Pb, Cr, Zn, Cd, and Sn; and 1.0 for Cu. The TOC content was determined on 1 g of dry homogenised sediment sample using the Walkley-Black wet oxidation method after decarbonisation of the sediment with excess HCl (Schumacher 2002; Botwe et al. 2017a). The carbonate content was analysed by treating sediment samples with excess HCl (1 M), followed by titration with 0.5 M NaOH (Kelderman and Osman 2007). The sediment pH was determined on 10 g portions of homogenised sediment samples after the addition of 25 mL deionised water following Palma and Mecozzi (2007).

To characterise stability of the Tema Harbour sediment against small changes in pH, the buffer intensities of the sediments were evaluated following the procedure described by Calmano (1988). Briefly, 5 g portions of sediment were suspended in 50 mL deionised water (10% sediment suspensions) in acid-washed 50 mL polypropylene tubes and shaken on an end-to-end shaker for 1 h, after which the pH of the supernatant solutions (pH_o) were measured. This procedure was repeated using 50 mL of 0.1 M HCl and the corresponding pH (pH_x) measured. The buffer intensity (δpH) was determined using Eq. (1) (Calmano 1988):

$$\delta pH = pH_o - pH_x \qquad (1)$$

The buffer intensities of the sediments were characterised as follows: $\delta pH < 2$ (strongly buffered), $\delta pH = 2\text{-}4$ (intermediate), and $\delta pH > 4$ (poorly buffered) (Calmano 1988).

For quality assurance and control, chemicals, solvents and reagents used in the extraction were of trace metal analysis grade. All containers used were either new or thoroughly cleaned by washing with detergent, soaking in 10% HNO_3 solution overnight and rinsing with deionised water. A sediment certified reference material (BCR-701) and blanks were analysed with the sediment samples. Sediment samples were analysed in triplicates. Metal recoveries ranged from 69.2-109% for the investigated metals, with relative standard deviations (RSD) in the range of 2.2-12.5 %. The reported metal concentrations were corrected for recoveries, using their respective mean recoveries from the certified reference materials.

5.2.2. Metal enrichment and contamination level

Metal enrichment factors (EFs) were derived for the measured metals to characterise their enrichment in the Tema Harbour sediments based on Eq. (2) (Dung et al. 2013; Mahu et al. 2015):

$$EF = [M/Al]_{Sample}/[M/Al]_{Crust} \qquad (2)$$

where $(M/Al)_{Sample}$ is the metal-aluminium ratio for the Tema Harbour sediment sample and $(M/Al)_{Crust}$ is the metal-aluminium ratio for the continental crust. $EF \leq 1$ indicates no enrichment, $1 < EF \leq 3$: minor enrichment, $3 < EF \leq 5$: moderate enrichment, $5 < EF \leq 10$: moderately severe enrichment, $10 < EF \leq 25$: severe enrichment, $25 < EF \leq 50$: very severe enrichment, and $EF > 50$: extremely severe enrichment (Dung et al. 2013).

Geo-accumulation indices (I_{geo}) were evaluated to assess the extent of metal contamination in the Tema Harbour sediments using Eq. (3) (Dung et al. 2013; Mahu et al. 2015):

$$I_{geo} = Log_2 (M_s/1.5M_b) \qquad (3)$$

where M_s is the metal concentration in the Tema Harbour sediment, M_b is its average crustal concentration and 1.5 is introduced as a background matrix correction due to lithogenic effects (Addo et al. 2011; Iqbal et al. 2013; Mahu et al. 2015). $I_{geo} < 0$: unpolluted; $0 \leq I_{geo} < 1$: unpolluted to moderately polluted; $1 \leq I_{geo} < 2$: moderately polluted; $2 \leq I_{geo}$ 3: moderately to highly polluted; $3 \leq I_{geo} < 4$: heavily polluted; $4 \leq I_{geo} < 5$: highly to very highly polluted; $I_{geo} \geq 5$: very heavily polluted (Dung et al. 2013).

In the evaluation of the EFs and I_{geo}, the average crustal concentrations of the metals (Taylor 1964) were used as proxies for the background concentration (Table 5.1), since pre-impacted or deep-core metal concentration data are not available for the Tema Harbour.

5.2.3. Data treatment

Pearson's product-moment correlations were performed to determine linear relationships between the measured sediment parameters using the Statistical Package for Social Sciences (SPSS 16.0) software. Descriptive statistics (means, standard deviations and coefficients of variation) were performed in Microsoft Excel 2007 for Windows at the 95% confidence interval. The spatial distribution patterns of metal concentrations in the Tema Harbour sediments were mapped in ArcGIS 10.2.2. One-way analysis of variance (ANOVA) with multiple comparison (Holm-Sidak and Kruskal-Wallis) tests was performed in SigmaPlot 11.0 to determine spatial differences in sediment metal concentrations in the Tema Harbour.

5.3. Results

5.3.1. Physicochemical characteristics of the Tema Harbour sediments

The sediment physicochemical parameters and *in situ* redox potential (E_h) at the sediment-water interface in the Tema Harbour are shown in Table 5.1. The Tema Harbour sediments were composed mainly of the silt-clay (<63 μm) fraction with a mean composition of 70 ($\pm$ 4)%. The sediment TOC and carbonate contents were in the ranges of 1.1-6.0% and 8.8-10.6%, respectively. The Tema Harbour sediments were slightly alkaline (pH range of 7.3-8.3) with buffer intensities in the range of 1.0-2.7, while the bottom water was anoxic with E_h values ranging from -70 to -350 mV. Sediments from the Canoe Basin were mostly anoxic. The silt-clay content, E_h and buffer intensity varied significantly across the Tema Harbour compartments. On the contrary, the pH, TOC and carbonate contents did not vary significantly across the Tema Harbour compartments. Table 5.2 shows that the silt-clay content correlated significantly with the Mn ($r = 0.52$, $p < 0.05$), Ni ($r = 0.52$, $p < 0.05$), Cr ($r = 0.58$, $p < 0.01$), Cu ($r = 0.60$, $p < 0.01$), Zn ($r = 0.75$, $p < 0.01$), As ($r = 0.64$, $p < 0.01$) and Sn ($r = 0.50$, $p < 0.05$) concentrations in the Tema Harbour sediments. The TOC content correlated with the Mn ($r = 0.48$, $p < 0.05$), Ni ($r = 0.46$, $p < 0.05$) and Cu ($r = 0.49$, $p < 0.05$) concentrations. The carbonate content correlated with the Hg concentration ($r = 0.47$, $p < 0.05$). The buffer intensity correlated with the Cr ($r = 0.85$, $p < 0.01$), Zn ($r = 0.71$, $p < 0.01$), As ($r = 0.83$, $p < 0.01$) and Sn ($r = 0.65$, $p < 0.01$) concentrations as well as with the silt-clay content ($r = 0.55$, $p < 0.01$). Sediment pH did not correlate significantly with the metal concentrations in the Tema Harbour sediments.

5.3.2. Metal concentrations and distribution patterns in sediments of Tema Harbour

The total metal concentrations in the Tema Harbour sediments are presented in Table 5.1. Overall, the mean sediment metal concentrations were: Al (17930 ± 780 mg.kg^{-1} dw), Mn (210 ± 7 mg.kg^{-1} dw), Ni (23 ± 1 mg.kg^{-1} dw), Cr (72 ± 3 mg.kg^{-1} dw), Pb (88 ± 13 mg.kg^{-1} dw), Cu (102 ± 16 mg.kg^{-1} dw), Zn (260 ± 40 mg.kg^{-1} dw), As (14.2 ± 0.6 mg.kg^{-1} dw), Cd (8.1 ± 0.6 mg.kg^{-1} dw), Hg (1.5 ± 0.3 mg.kg^{-1} dw) and Sn (8.2 ± 1.7 mg.kg^{-1} dw). Significant positive correlations were observed among most of the sediment metal concentrations (Table 5.2). The Mn and Ni concentrations showed a perfect correlation ($r = 1.00$, $p < 0.01$) and both correlated significantly with all the other metals, except Pb, Hg and Sn. Cd exhibited significant correlations with all the metals except Cr. Pb exhibited a significant correlation with only Al ($r = 0.60$, $p < 0.01$), whereas Hg did not exhibit any significant correlation with all the other metals investigated.

The spatial distribution patterns of the metals are shown in Fig. 5.1. Sediment Al concentrations were relatively higher in the Inner Fishing Harbour and the Main Harbour entrance stations. Mn concentrations were relatively higher in the Fishing Harbour, Canoe Basin, and the entrance and mid-section of the Main Harbour. Relatively higher Ni concentrations occurred in the north-eastern corner of the Main Harbour. Sediments from the Fishing Harbour and the Canoe Basin contained relatively higher Cr concentrations. Sediment concentrations of As were relatively higher in the Inner Fishing Harbour and the Canoe Basin. Relatively higher sediment Cd concentrations occurred in the Inner Fishing Harbour. Sediments from the Canoe Basin and the Fishing Harbour contained relatively higher concent-

Table 5.1 Sediment metal concentrations and physicochemical parameters measured in the Tema Harbour, Ghana, their average crustal concentrations and metal concentrations in Tema Harbour sediments from previous studies in 2000 and 2010-2011.

Harbour area	Sampling station	Metal concentration (mg.kg⁻¹ dry weight)											Physicochemical parameters					
		Al	Mn	Ni	Pb	Cr	Cu	Zn	As	Cd	Hg	Sn	Silt-clay (%)	TOC (%)	CaCO₃ (%)	pH	Buffer intensity	E$_h$ (mV)
MH	1	17980	194	21.0	30.5	69.2	81.1	205	14.9	7.5	2.6	23.9	67	4.5	9.5	8.0	1.9	-110
	2	14170	165	21.3	49.3	57.8	77.2	160	10.0	5.1	2.1	28.8	67	5.4	9.4	8.0	1.4	-130
	3	14100	160	21.1	45.3	56.6	75.1	155	9.6	7.5	1.4	10.6	67	4.7	9.7	7.7	1.6	-100
	4	27500	270	26.2	36.3	73.9	79.4	185	15.0	8.7	1.9	12.7	68	5.8	9.0	7.6	1.0	-90
	5	19280	210	22.8	38.1	66.3	43.3	124	14.5	8.1	0.4	0.8	39	6.0	9.0	8.0	1.0	-90
	6	17640	195	21.6	40.0	63.0	44.8	128	14.2	17.7	0.4	0.6	70	2.4	10.6	8.0	1.3	-90
	7	14170	183	21.4	57.6	58.9	42.5	164	10.0	4.3	0.4	0.7	53	1.1	9.8	7.6	1.3	-80
	8	17890	177	24.7	33.1	63.3	34.2	77	14.0	8.0	0.3	0.5	57	2.2	9.7	7.3	1.4	-90
	9	19070	187	23.2	56.8	60.9	46.5	110	9.8	9.2	0.6	0.7	84	5.4	9.5	8.0	1.3	-80
	10	16730	198	20.0	68.4	58.8	44.5	89	14.9	6.0	0.3	0.7	62	2.2	9.3	8.1	1.3	-70
	11	16220	198	21.9	63.3	70.2	30.9	137	14.1	5.4	0.4	0.6	52	4.7	9.5	7.8	1.3	-80
	12	18100	187	26.1	57.8	55.8	36.2	123	12.6	7.7	0.8	0.9	71	4.1	9.2	7.6	1.3	-90
	13	20260	223	35.9	79.9	77.6	78.7	208	15.6	5.8	0.9	14.7	66	5.8	9.6	7.8	1.4	-90
	14	16840	250	27.6	73.5	71.6	66.3	170	15.3	9.4	0.7	11.3	80	4.5	9.2	8.3	1.3	-90
CB	15	15810	240	18.7	230	71.3	196	576	17.0	6.9	0.8	7.8	85	4.7	9.4	8.2	2.7	-350
	16	15420	230	20.9	206	77.3	180	561	14.0	8.2	4.3	6.8	80	4.2	8.8	7.6	2.1	-260
	17	13870	210	20.1	216	65.6	168	532	15.8	5.9	4.5	9.0	82	4.0	9.4	7.5	2.7	-240
IFH	18	19050	270	22.6	130	110	264	518	21.3	9.2	1.6	12.9	89	5.8	9.0	7.8	1.8	-130
	19	26710	260	30.4	122	106	220	500	18.7	12.7	2.2	14.0	87	5.4	9.8	7.6	1.7	-140
OFH	20	18780	250	21.9	107	89.4	170	395	15.1	8.7	2.6	7.9	92	5.3	9.4	7.9	2.3	-150
	21	16970	220	19.9	97.6	95.4	163	240	13.7	7.1	2.2	6.4	93	5.5	9.2	7.8	1.7	-160
	Mean	17930	210	23	88	72	102	260	14.2	8.1	1.5	8.2	70	4.5	9.4	-	1.6	-
	*ACC	82300	950	75	12.5	100	55	70	1.8	0.2	0.08	2						
	a Year 2000	-	-	-	51.5	209	-	-	9.2	0.2	4.4	-						
	b Year 2010-2011	-	41.7	34.2	-	nd	25.8	28.5	1.5	13.4	-	-						

MH = Main Harbour; CB = Canoe Basin; IFH = Inner Fishing Harbour; OFH = Outer Fishing Harbour; *ACC = average crustal concentration (Taylor 1964)

a JICA (http://open_jicareport.jica.go.jp/pdf/11681632_03.pdf); b Nyarko et al. (2014); nd = not detected; '-' = not applicable or not measured.

Table 5.2 Correlation matrix of measured metals and physicochemical characteristics of surface sediments from Tema Harbour, Ghana (2-tailed; $n = 21$)

	Al	Mn	Ni	Pb	Cr	Cu	Zn	As	Cd	Hg	Sn	Silt-clay	TOC	Carbonate	pH
Mn	0.61**														
Ni	0.61**	1.00**													
Pb	0.60**	0.30	0.30												
Cr	-0.20	0.44*	0.44**	-0.22											
Cu	0.48*	0.78**	0.78**	0.23	0.36										
Zn	0.16	0.67**	0.67**	0.09	0.76**	0.81**									
As	0.04	0.61**	0.61**	-0.12	0.91**	0.61**	0.92**								
Cd	0.44*	079**	0.79**	0.23	0.45*	0.74**	0.69**	0.61**							
Hg	0.42	0.27	027	0.15	-0.10	0.26	0.12	0.06	0.28						
Sn	-0.05	0.27	0.27	-0.19	0.58**	0.32	0.61**	0.67**	0.18	-0.08					
Silt-clay	0.12	0.52*	0.52*	-0.03	0.58**	0.60**	0.75**	0.64**	0.36	0.27	0.50*				
TOC	0.42	0.48*	0.46*	0.29	0.12	0.49*	0.40	0.27	0.27	-0.01	0.27	0.34			
Carbonate	-0.07	-0.40	-0.40	0.07	-0.29	-0.23	-0.29	-0.27	-0.20	0.47*	-0.34	-0.13	-0.48		
pH	-0.16	0.10	0.10	-0.17	-0.05	-0.08	-0.08	-0.10	-0.04	0.11	-0.31	0.10	0.20	-0.02	
BI	-0.31	0.25	0.25	-0.37	0.85**	0.29	0.71**	0.83**	0.35	-0.11	0.65**	0.60**	0.07	-0.09	-0.03

*Correlation is significant at the 0.05 level.

** Correlation is significant at the 0.01 level.

BI = Buffer intensity

rations of Pb, Cu, Zn and Hg, while sediments from the north-eastern corner of the Main Harbour contained higher concentrations of Sn. Analysis of variance (ANOVA) revealed that the Mn, Ni, Cr, Cu, Zn, As and Sn concentrations differed significantly ($p < 0.05$) among the Main Harbour, the Inner Fishing Harbour, the Outer Fishing Harbour and the Canoe Basin.

5.3.3. Metal fractionation, enrichment and pollution in Tema Harbour sediments

The distribution of metals over the acid-soluble, reducible, oxidisable and residual phases of sediments from the Tema Harbour are shown in Fig. 5.2. At all the sampling stations, Cd occurred predominantly in the acid-soluble phase (70-85%), and was divided nearly equally among the reducible (5-10%), oxidisable (5-12%) and residual (5-10%) phases. Hg also occurred mainly in the acid-soluble phase (52-67) with relatively higher fractions in the residual phase (15-42%) than in the reducible (0-15%) and oxidisable (0-14%) phases. At all the sampling stations, Al (73-83%), Mn (70-80%), Ni (62-72%), Pb (50-65%), Cr (60-76%), Cu (57-72%), As (52-63%) and Sn (50-100%) were mainly present in the residual phase. For most of these metals, the fractions in the acid-soluble phase were low, i.e. Al (4-8%), Mn (8-17%), Ni (8-15%), Pb (5-10%), Cr (3-7%), Cu (3-7%) and As (8-15%). In the case of Sn, no appreciable fraction was present in the acid-soluble phase at stations S5-S12 in the Main Harbour, although their fraction was appreciable (17-28%) at the remaining sampling stations. For Zn, both the reducible (38-45%) and residual (35-42%) phases were the most important at all the sampling stations, while the oxidisable phase was the least important. Appreciable fractions of Zn (12-18%) were also present in the acid-soluble phase.

Table 5.3 shows the enrichment factors (EFs) of the investigated metals in the Tema Harbour sediments. The results show relatively higher enrichment of Cd over the other metals. The Cd concentrations in the Tema Harbour surface sediments exceeded the average crustal concentration by over an order of magnitude. The concentrations of Cd, Hg, Pb and As in the Tema Harbour sediments were 12-41, 2-33, 1-10 and 2-5 times their average crustal concentrations, respectively. The concentrations of Cu and Zn in the Tema Harbour sediments were elevated above their average crustal concentrations only in the Canoe Basin and the Fishing Harbour, while the sediment Sn concentrations exceeded the average crustal concentration only at stations S1-S4, S13 and S14 in the north-eastern corner of the Main Harbour, the Canoe Basin and the Fishing Harbour. The concentrations of Mn, Ni and Cr in the Tema Harbour sediments were about an order of magnitude lower than their average

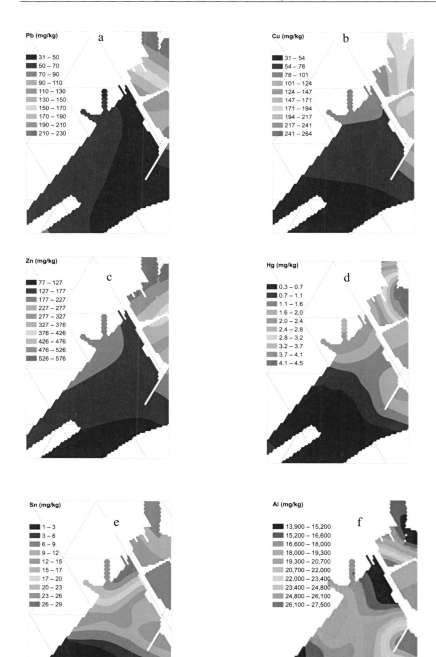

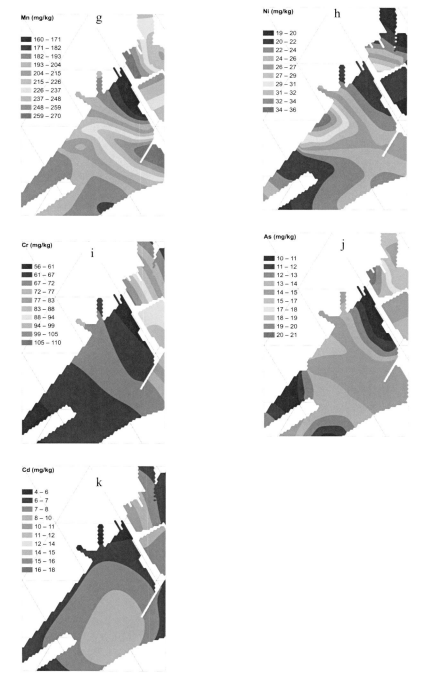

Fig. 5.1 Spatial distribution patterns of (a) Pb, (b) Cu, (c) Zn, (d) Hg, (e) Sn, (f) Al, (g) Mn, (h) Ni, (i) Cr, (j) As and (k) Cd in surface sediments of Tema Harbour

crustal concentrations.

The potential elevation of metal concentrations in the Tema Harbour sediments was confirmed by the geo-accumulation indices (Table 5.3). The geo-accumulation indices categorised all the Tema Harbour sediments as heavily polluted with respect to Cd, but unpolluted with respect to the Mn, Ni and Cr concentrations. Sediments of the Canoe Basin (stations S15-S17) were categorised, based on the geo-accumulation index, as polluted with respect to the Pb and Zn concentrations, those of the Fishing Harbours (stations S18-S21) as polluted with respect to the Hg and Cu concentrations, and those of stations S1-S4 in the north-eastern corner of the Main Harbour as polluted with respect to the Hg concentrations.

5.4. Discussion

5.4.1. Distribution and potential sources of metals in Tema Harbour surface sediments

This study showed high enrichment of Cd, Hg and As in the Tema Harbour surface sediments (EFs > 10), which indicates that Cd, Hg and As contamination in the Tema Harbour sediments is mainly of anthropogenic origin (Chatterjee et al. 2007; Addo et al. 2011; Mahu et al. 2015). Similarly, the high Pb enrichment in all the sediment samples (except for station S4) suggests significant anthropogenic inputs of Pb in the Tema Harbour sediments. The EFs of Cu, Zn and Sn also point to anthropogenic contributions of these metals in various parts of the Tema Harbour (Table 5.3). The low enrichment of Mn, Ni and Cr in the sediment samples (EFs < 10) indicate that the sources of these metals in the Tema Harbour sediments were mainly from natural or lithogenic sources, such as weathering and chemical leaching of the bedrock. The metal enrichment in the Tema Harbour sediments is supported by the geo-accumulation indices (I_{geo}) (Table 5.3), which indicate that the Tema Harbour sediments are potentially contaminated with respect to their Pb, Cu, Zn, As, Cd, Hg and Sn concentrations, while they are potentially uncontaminated with respect to the Mn, Ni and Cr concentrations.

Potential anthropogenic sources of Pb, Cu, Zn, As, Cd, Hg and Sn contamination in the Tema Harbour sediments are varied and may relate to bunkering, fueling, discharge of bilge, and scraping and painting of vessels as well as vehicular traffic, industrial and urban effluent discharges (Lepland et al. 2010; Mestres et al. 2010; Schipper et al. 2010; Nyarko et al. 2014). The use of antifouling paints on ships and fishing vessels, for example, is an important source

Table 5.3 Enrichment factors and geo-accumulation indices of measured metals in surface sediments from Tema Harbour, Ghana

Sampling station	Enrichment Factors (EFs)										Geo-accumulation indices (I_{geo})									
	Mn	Ni	Pb	Cr	Cu	Zn	As	Cd	Hg	Sn	Mn	Ni	Pb	Cr	Cu	Zn	As	Cd	Hg	Sn
S1	0.9	1.3	11.2	3.2	6.7	13.4	37.9	172	149	54.7	-2.9	-2.4	0.7	-1.1	0.0	1.0	2.5	4.6	4.4	3.0
S2	1.0	1.6	22.9	3.4	8.2	13.3	32.3	148	153	83.6	-3.1	-2.4	1.4	-1.4	-0.1	0.6	1.9	4.1	4.1	3.3
S3	1.0	1.6	21.2	3.3	8.0	12.9	31.1	219	102	30.9	-3.2	-2.4	1.3	-1.4	-0.1	0.6	1.8	4.6	3.5	1.8
S4	0.9	1.0	8.7	2.2	4.3	7.9	24.9	130	71.1	19.0	-2.4	-2.1	1.0	-1.0	-0.1	0.8	2.5	4.9	4.0	2.1
S5	0.9	1.3	13.0	2.8	3.4	7.6	34.4	173	21.3	1.7	-2.8	-2.3	1.0	-1.2	-0.9	0.2	2.4	4.8	1.7	-1.9
S6	1.0	1.3	14.9	2.9	3.8	8.5	36.8	413	23.3	1.4	-2.9	-2.4	1.1	-1.3	-0.9	0.3	2.4	5.9	1.7	-2.3
S7	1.1	1.7	26.8	3.4	4.5	13.6	32.3	125	29.0	2.0	-3.0	-2.4	1.6	-1.3	-1.0	0.6	1.9	3.8	1.7	-2.1
S8	0.9	1.5	12.2	2.9	2.9	5.1	35.8	184	17.3	1.2	-3.0	-2.2	0.8	-1.2	-1.3	-0.4	2.4	4.7	1.3	-2.6
S9	0.8	1.3	19.6	2.6	3.6	6.8	23.5	199	32.4	1.5	-2.9	-2.3	1.6	-1.3	-0.8	0.1	1.9	4.9	2.3	-2.1
S10	1.0	1.3	26.9	2.9	4.0	6.2	40.7	148	18.4	1.7	-2.8	-2.5	1.9	-1.4	-0.9	-0.2	2.5	4.3	1.3	-2.1
S11	1.1	1.5	25.7	3.6	2.9	9.9	39.7	137	25.4	1.5	-2.8	-2.4	1.8	-1.1	-1.4	0.4	2.4	4.2	1.7	-2.3
S12	0.9	1.6	21.0	2.5	3.0	8.0	31.8	175	45.5	2.0	-2.9	-2.1	1.6	-1.4	-1.2	0.2	2.2	4.7	2.7	-1.7
S13	1.0	1.9	26.0	3.2	5.8	12.1	35.2	118	45.7	29.9	-2.7	-1.6	2.1	-1.0	-0.1	1.0	2.5	4.3	2.9	2.3
S14	1.3	1.8	28.7	3.5	5.9	11.9	41.5	230	42.8	27.6	-2.5	-2.0	2.0	-1.1	-0.3	0.7	2.5	5.0	2.5	1.9
S15	1.3	1.3	95.8	3.7	18.6	42.8	49.2	180	52.1	20.3	-2.6	-2.6	3.6	-1.1	1.2	2.5	2.7	4.5	2.7	1.4
S16	1.3	1.5	88.0	4.1	17.5	42.8	41.5	219	287	18.1	-2.6	-2.4	3.5	-1.0	1.1	2.4	2.4	4.8	5.2	1.2
S17	1.3	1.6	103	3.9	18.1	45.1	52.1	175	334	26.7	-2.8	-2.5	3.5	-1.2	1.0	2.3	2.5	4.3	5.2	1.6
S18	0.2	1.3	44.9	4.8	20.7	32.0	51.1	199	86.4	27.9	-2.4	-2.3	2.8	-0.4	1.7	2.3	3.0	4.9	3.7	2.1
S19	0.8	1.2	30.1	3.3	12.3	22.0	32.0	196	84.7	21.6	-2.5	-1.9	2.7	-0.5	1.4	2.3	2.8	5.4	4.2	2.2
S20	1.2	1.3	37.5	3.9	13.5	24.7	36.8	191	142	17.3	-2.5	-2.4	2.5	-0.7	1.0	1.9	2.5	4.9	4.4	1.4
S21	1.1	1.3	37.9	4.6	14.4	16.6	36.9	172	133	15.5	-2.7	-2.5	2.4	-0.7	1.0	1.2	2.3	4.6	4.2	1.1

Main Harbour (stations S1-S14); Canoe Basin (stations S15-S17); Inner Fishing Harbour (stations S18 and S19); Outer Fishing Harbour (stations 20 and 21). For the interpretations of the EF I_{geo} values, see Section 5.2.2.

of Cu and Sn in harbour sediments (Castro et al. 2012; Nyarko et al. 2011). There were significant correlations among the metals (Table 5.2), which can be attributed to common sources (Nyarko et al. 2014; Qu and Kelderman 2001) or similar distribution patterns (Aloupi and Angelidis 2001) of the metals in the Tema Harbour sediments.

Different proximities of sampling stations to local contamination sources in the Tema Harbour may account for the differences in the sediment metal concentrations across the harbour. For example, the Tema Harbour has two dry docks located close to stations S1 and S2, where activities such as sandblasting, high pressure water cleaning, scraping and painting are carried out. Moreover, station S4 is close to the Oil Berth, where oil tankers berth and carry out bunkering activities. Thus, the proximity of the north-eastern stations to the dry docks and the Oil Berth in the Main Harbour may have resulted in the relatively higher Cu, Hg and Sn concentrations present in the sediments from this area. Higher concentrations of Pb, Cu, Zn and Hg occurred in the Canoe Basin and the Fishing Harbour, which could be due to oil spills during refuelling of fishing vessels in these areas.

Moreover, the Main Harbour is frequently dredged, whereas the Fishing Harbour and Canoe Basin are not. Dredging in the Main Harbour may potentially remove contaminated surficial sediments and result in reduced metal levels, whereas the accumulation of sediments in the Canoe Basin and the Fishing Harbour may result in elevated concentrations of associated metals. The resuspension and redistribution of bottom sediment due to e.g. tidal currents, dredging and ship movements (Lepland et al. 2010) may also contribute to the spatial variations in the sediment metal concentrations in the Tema Harbour.

Since bulk sediments were analysed in this study, a correlation analysis was performed to examine the associations between the metal concentrations and sediment characteristics. Significant correlations were observed between the silt-clay content and the Cu, Zn, Sn, Mn, Ni, Cr and As concentrations in the Tema Harbour sediments (Table 5.2), which can be attributed to the high affinity of metals for the fine fractions of sediments owing to their large surface areas (Aloupi and Angelidis 2001; Kelderman and Osman 2007; Lepland et al. 2010; Nyarko et al. 2011). The distribution patterns of Cu, Zn, Sn, Mn, Ni, Cr and As in the Tema Harbour sediments (Fig. 5.1) may, therefore, be influenced by the differences in their silt-clay content. As a geochemical substrate for metals (Aloupi and Angelidis 2001), TOC may also play a role in the metal distribution, which may explain its correlations with Mn, Ni and Cu (Table 5.2).

Carbonate generally has a low capacity to adsorb and retain metals and is, therefore, considered as a diluent of metal concentrations in sediments (Horowitz 1985; Aloupi and Angelidis 2001). Thus, Aloupi and Angelidis (2001) found an inverse relationship between the metal and carbonate concentration of the Mytilene Harbour sediments. The Tema Harbour sediments had a nearly uniform carbonate content (Table 5.1), exhibiting generally poor correlations with the metal concentrations in the harbour sediments (Table 5.2). Al is known to distribute nearly equally between the fine and coarse fractions of a sediment (Aloupi and Angelidis 2001), which may explain the observed poor correlation between the Al concentration and the silt-caly fraction of the habour sediments (Table 5.2).

The mean metal concentrations from this study were compared with previous studies in 2000 by JICA (Japan International Co-operation Agency) and 2010-2011 by Nyarko et al. (2014) (Table 5.1). Table 5.1 shows that the current mean Pb, As and Cd concentrations are higher than the levels in 2000, whereas the current concentrations of Cr and Hg are lower than the 2000 levels. Moreover, the current mean levels of Mn, Cr, Cu, Zn and As are higher than the levels in 2010-2011. The mean reported concentrations of Ni and Cd in 2010-2011 were, however, higher than the current levels. These comparisons did not reveal a progressive increase in metal contamination of Tema Harbour sediments since the year 2000.

5.4.2. Metal fractionation in the Tema Harbour sediments

The negative E_h values of the bottom water (Table 5.1) indicate anoxic bottom conditions in the Tema Harbour (Matijević et al. 2007), which has also been observed in previous studies (Botwe et al. 2017a; Botwe et al. 2017b). Metal fractionation in the Tema Harbour sediments (Fig. 5.2) showed that most of the metals (Al, Mn, Ni, Pb, Cr, Cu, As and Sn) were predominantly present in the residual phase, where they are incorporated into the crystal lattice (Calmano et al. 1993; Kelderman and Osman 2007). Among the non-residual fractions, the oxidisable phase was most important for Pb, Cr, Cu and As. This indicates that complexation with organic matter and precipitation with sulphides were the main mechanisms for the immobilisation of Pb, Cr, Cu and As into the Tema Harbour sediments, which is the case for most metals under anoxic conditions (Caille et al. 2003; Ho et al. 2012a).

Mn and Sn were mainly present in the acid-soluble phase, while Al and Ni were distributed nearly equally between the non-residual phases. The fractionation of Zn showed that the reducible phase was most important. This suggests that co-precipitation with oxides or hydroxides of Fe and Mn was a main mechanism for Zn immobilisation into the Tema Harbour sediments. Cd (70-85%) and Hg (52-67%) were predominantly associated with the acid-soluble phase. In this phase, Hg may associate with carbonates, which may partly explain the observed correlation between Hg and carbonate (Table 5.2).

5.4.3. Ecological implications of metal contamination in Tema Harbour sediments

5.4.3.1. Ecotoxicological implications

Metal contamination in sediments may cause toxicity in sediment-dwelling organisms (Long et al. 1995; Long et al. 2006). The effects-range low (ERL) and effects-range median (ERM) sediment quality guidelines (Long et al. 1995) were used to characterise the potential toxicity of the Tema Harbour sediments due to their metal (Ni, Pb, Cr, Cu, Zn, As, Cd and Hg) contamination. Metal concentrations below the ERL, at or above the ERL but below the ERM, and at or above the ERM are associated with, respectively, rare, occasional, and frequent occurrence of toxic effects (Long et al. 1995). For each metal, two quotients namely the effects-range low qoutient (ERLQ) and effects-range median quotient (ERMQ) were derived by dividing the measured metal concentration by its corresponding ERL and ERM concentrations, respectively (Long et al. 2006). Within this approach, an ERLQ < 1 indicates toxic effects will rarely occur; an ERLQ ≥ 1 but ERMQ < 1 indicates toxic effects will occur occasionally; whereas an ERMQ ≥ 1 indicates toxic effects will occur frequently.

The ERLQ and ERMQ values (Table 5.4) indicate that Hg concentrations at most (67%) of the sampling stations may potentially cause frequent occurrence of toxic effects. Thus, Hg is of potential concern in the Tema Harbour, particularly in the north-eastern corner of the Main Harbour, the Canoe Basin and the Fishing Harbour. Frequent occurrence of toxic effects is also expected as a result of the sediment Zn concentrations in the Canoe Basin and the Fishing Harbour. There are also isolated areas in the Tema Harbour where the concentrations of Pb and Cd in the sediments may be associated with frequent occurrence of toxic effects. At most of the stations, however, there is a potential for occasional occurrence of toxic effects as a

Table 5.4 Calculated effects-range low quotients (ERLQ), effects-range median quotients (ERMQs) and mean ERM quotients (mERMQs) for measured metals in surface sediments from Tema Harbour, Ghana. The ERL and ERM values for the measured metals are also presented.

Harbour area	Sampling station	ERLQ								ERMQ								mERMQ
		Ni	Pb	Cr	Cu	Zn	As	Cd	Hg	Ni	Pb	Cr	Cu	Zn	As	Cd	Hg	
MH	S1	1.0	0.7	0.9	2.4	1.4	1.8	6.3	17.3	0.4	0.1	0.2	0.3	0.5	0.2	0.8	3.7	0.8
	S2	1.0	1.1	0.7	2.3	1.1	1.2	4.3	14.0	0.4	0.2	0.2	0.3	0.4	0.1	0.5	3.0	0.6
	S3	1.0	1.0	0.7	2.2	1.0	1.2	6.3	9.3	0.4	0.2	0.2	0.3	0.4	0.1	0.8	2.0	0.5
	S4	1.3	0.8	0.9	2.3	1.2	1.8	7.3	12.7	0.5	0.2	0.2	0.3	0.5	0.2	0.9	2.7	0.7
	S5	1.1	0.8	0.8	1.3	0.8	1.8	6.8	2.7	0.4	0.2	0.2	0.2	0.3	0.2	0.8	0.6	0.4
	S6	1.0	0.9	0.8	1.3	0.9	1.7	14.8	2.7	0.4	0.2	0.2	0.2	0.3	0.2	1.8	0.6	0.5
	S7	1.0	1.2	0.7	1.3	1.1	1.2	3.6	2.7	0.4	0.3	0.2	0.2	0.4	0.1	0.4	0.6	0.3
	S8	1.2	0.7	0.8	1.0	0.5	1.7	6.7	2.0	0.5	0.2	0.2	0.1	0.2	0.1	0.8	0.4	0.3
	S9	1.1	1.2	0.8	1.4	0.7	1.2	7.7	4.0	0.4	0.3	0.2	0.2	0.3	0.1	1.0	0.8	0.4
	S10	1.0	1.5	0.7	1.3	0.6	1.8	5.0	2.0	0.4	0.3	0.2	0.2	0.2	0.2	0.6	0.4	0.3
	S11	1.0	1.4	0.9	0.9	0.9	1.7	4.5	2.7	0.4	0.3	0.2	0.1	0.3	0.2	0.6	0.6	0.3
	S12	1.2	1.2	0.7	1.1	0.8	1.5	6.4	5.3	0.5	0.3	0.2	0.1	0.3	0.2	0.8	1.1	0.4
	S13	1.7	1.7	1.0	2.3	1.4	1.9	4.8	6.0	0.7	0.4	0.2	0.3	0.5	0.2	0.6	1.3	0.5
	S14	1.3	1.6	0.9	2.0	1.1	1.9	7.8	4.7	0.5	0.3	0.2	0.2	0.4	0.2	1.0	1.0	0.5
CB	S15	0.9	4.9	0.9	5.8	3.8	2.1	5.8	5.3	0.4	1.1	0.2	0.7	1.4	0.2	0.7	1.1	0.7
	S16	1.0	4.4	1.0	5.3	3.7	1.7	6.8	28.7	0.4	0.9	0.2	0.7	1.4	0.2	0.9	6.1	1.3
	S17	1.0	4.6	0.8	4.9	3.5	1.9	4.9	30.0	0.4	1.0	0.2	0.6	1.3	0.2	0.6	6.3	1.3
IFH	S18	1.1	2.8	1.4	7.8	3.5	2.6	7.7	10.7	0.4	0.6	0.3	1.0	1.3	0.3	1.0	2.3	0.9
	S19	1.5	2.6	1.3	6.5	3.3	2.3	10.6	14.7	0.6	0.6	0.3	0.8	1.2	0.3	1.3	3.1	1.0
OFH	S20	1.0	2.3	1.1	5.0	2.6	1.8	7.3	17.3	0.4	0.5	0.2	0.6	1.0	0.2	0.9	3.7	0.9
	S21	1.0	2.1	1.2	4.8	1.6	1.7	5.9	14.7	0.4	0.4	0.3	0.6	0.6	0.2	0.7	3.1	0.8
	ERL	20.9	46.7	81	34	150	8.2	1.2	0.15	-	-	-	-	-	-	-	-	-
	ERM	51.6	218	370	270	410	70	9.6	0.71	-	-	-	-	-	-	-	-	-

MH = Main Harbour; CB = Canoe Basin; IFH = Inner Fishing Harbour; OFH = Outer Fishing Harbour result of Ni, Pb, Cu, As and Cd concentrations in the sediments, while the Cr concentrations are not likely to cause toxic effects based on the ERLQ and ERMQ values (Table 5.4).

To characterise the overall potential sediment toxicity due to a mixture of metals, mean ERM quotients (mERMQs) were calculated by averaging the different metal ERMQs with the toxic effects (Long et al. 2006; Birch and Hutson 2009). The overall potential toxicity is classified as minimal if the mERMQ < 0.1, low if 0.1 ≤ mERMQ < 0.5, moderate if 0.5 ≤ mERMQ < 1.5 and high if mERMQ > 1.5 (Birch and Hutson 2009). The mERMQ values of the Tema Harbour sediments (Table 5.4) indicate that metal concentrations of the sediments from the north-eastern stations in the Main Harbour, the Canoe Basin and the Fishing Harbour may cause moderate toxicity to benthic organisms, while sediments from the remaining areas may be associated with low toxicity.

5.4.3.2. Ecological risk implications

Metal fractionation in sediments significantly determines their potential mobility and availability for uptake by benthic organisms (Calmano et al. 1993; Jain 2004; Pini et al. 2015). Metals associated with the acid-soluble phase have the greatest mobility and bioavailability potential and pose the greatest ecological risk, whereas those in the residual phase have the least mobility and bioavaility potential and pose the least ecological risk (Calmano et al. 1993; Jain 2004; Kelderman and Osman 2007; Iqbal et al. 2013; Ho et al. 2012a). A risk assessment code (RAC) has therefore been developed based on the fraction (%) of metal associated with the acid-soluble phase to characterise the potential risk of sediment-associated metals entering the food chain as follows (Jain 2004): acid-soluble fraction <1 % indicates no risk, 1-10 % indicates low risk, 11-30 % indicates medium risk, 31-50% indicates high risk, and >50 % indicates very high risk.

The phase distributions of the investigated metals in the Tema Harbour sediments (Fig. 5.2) show large fractions of Cd (70-85%) and Hg (52-67) in the acid-soluble phase, indicating that Cd and Hg contamination in the Tema Harbour sediments may pose very high ecological risks. Sn may pose medium risk at stations 1-4 in the Main Harbour, the Canoe Basin (stations 15-17) and the Fishing Harbour (stations 18-21), where the acid-soluble fraction ranged between 17 and 28%, but poses no risk at the remaining stations. The fractions of Mn (8-17%), Ni (8-15%), Zn (12-18%) and As (8-15%) in the acid-soluble phase indicate these

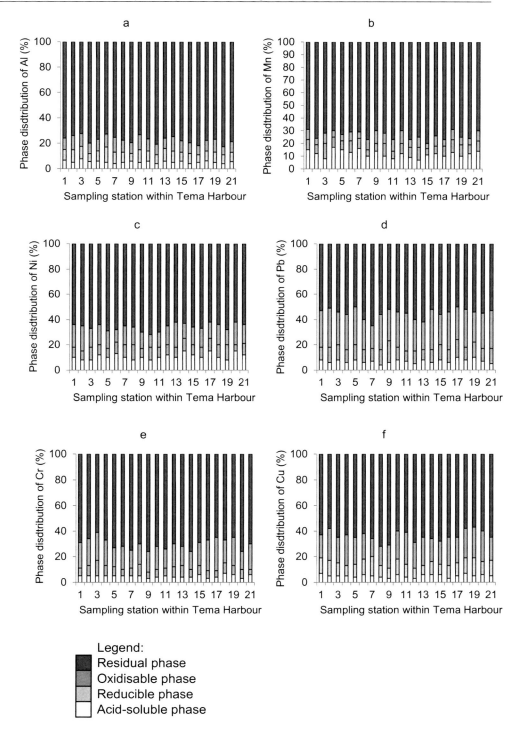

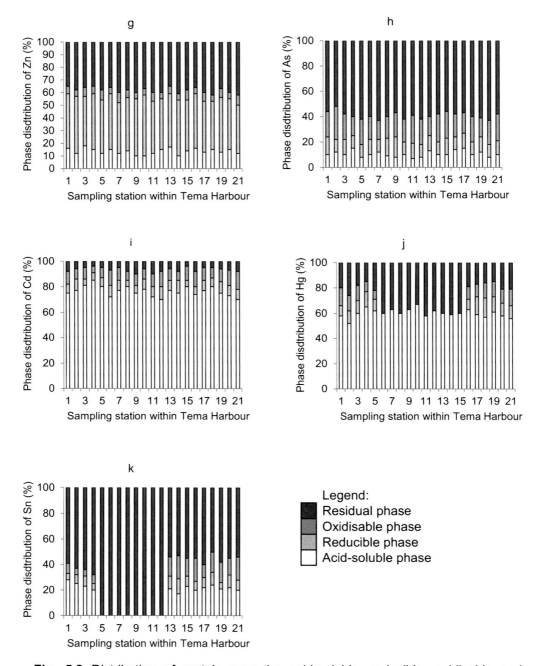

Fig. 5.2 Distribution of metals over the acid-soluble, reducible, oxidisable and residual phases of Tema Harbour sediments: (a) Al, (b) Mn, (c) Ni, (d) Pb, (e) Cr, (f) Cu, (g) Zn, (h) As, (i) Cd, (j) Hg, and (k) Sn

metals pose low-medium risks, while those of Pb (5-10%), Cr (3-7%) and Cu (3-7%) present low risk. The non-residual or labile (i.e. acid-soluble + reducible + oxidisable) fractions, however, constitute a pool of potentially bioavailable metals due to their potential to undergo changes under varying environmental conditions such as pH, E_h and resuspension of bottom sediments (Calmano et al. 1993; Kelderman and Osman 2007; De Jonge et al. 2012; Dung et al. 2013; Hamzeh et al. 2014; Pini et al. 2015). The considerable fractions of Mn (20-30%), Ni (28-38%), Pb (35-50%), Cr (24-40%), Cu (28-43%), As (37-48%) and Sn (0-50%) present in the labile phase, therefore, have implications for their potential ecological risks.

5.4.4. Chemical stability of Tema Harbour sediments

The pH range of the Tema Harbour sediments (7.3-8.3; Table 5.1) indicates that all the investigated sediments were slightly alkaline. The buffer intensities of the Tema Harbour sediments ranged from intermediate to strong (Table 5.1). Thus, in general, the Tema Harbour sediments are potentially stable against small changes in pH, with the potential to lower the risk of metal remobilisation and exposure due to acidification. A reduction in pH (acidifcation) can substantially increase the risk of metal exposure to organisms via remobilisation of sediment-associated metals (Calmano 1988; Calmano et al. 1993; Kelderman and Osman 2007; Ho et al. 2012a). The potential ecological risk of metal remobilisation due to acidification may be more pronounced in weakly buffered sediments, since they have a low capacity to resist slight changes in pH.

The carbonate content is a major determinant of the buffer intensity of sediments: sediments with carbonate contents >10% are well buffered against acidification over a wide range (Calmano 1988). The carbonate contents of the Tema Harbour sediments were close to 10 (Table 5.1) and exhibited no significant correlation with the buffer intensity (Table 5.2). The observed significant correlations between buffer intensity and Cr, Zn, As and Sn concentrations in the Tema Harbour sediments indicate that increasing buffer intensity enhances the immobilisation and accumulation of these metals in the harbour sediments. The significant correlation between buffer intensity and silt-clay content (Table 5.2) suggests that the silt-clay content has a positive influence on the buffer intensity of the Tema Harbour sediments, possibly due to the effect of alumino silicates, which are another major determinant of the sediment buffer intensity (Calmano 1988).

5.5. Conclusions

The metal (Al, Mn, Ni, Pb, Cr, Cu, Zn, As, Cd, Hg and Sn) distribution, fractionation and their ecological implications of surface sediments from the Tema Harbour (Greater Accra, Ghana) have been investigated. Sediment concentrations of Mn, Ni, Cr, Cu, Zn, As and Sn differed significantly across the Tema Harbour. Cd, Hg, Pb, Cu, Zn, As and Sn were enriched in the Tema Harbour sediments and may have been derived mainly from anthropogenic sources. As the Tema Harbour is located in an industrial area, potential anthropogenic sources of metals in the harbour include industrial effluent discharges as well as vehicular emissions. Moreover, metal contamination in the Tema Harbour may originate from shipping and fishing activities such as the use of antifouling paints on ships and fishing vessels, oil spills, discharge of bilge, and scraping and painting of vessels. On the other hand, Mn, Ni and Cr were not enriched in the Team Harbour sediments and may, therefore, be of lithogenic origin. Al, Mn, Ni, Pb, Cr, Cu, As and Sn were present mainly in the residual phase, Cd and Hg associated mainly with the exchangeable phase, while the reducible phase was most important for Zn. Based on the metal fractionation in the sediments, Cd and Hg may pose high potential risks of entering the food chain; Mn, Ni, Zn, As and Sn may pose low-medium potential risks of entering the food chain, while Pb, Cr, and Cu pose low potential risks of entering the food chain. A screenining-level ecotoxicological evaluation of the sediment metal concentrations with reference to biological effect-based sediment quality guidelines indicated that Hg may be associated with high toxicity and should, therefore, be of potential concern. Pb, Cu, Zn, As and Cd concentrations may also induce appreciable toxicity.

References

Addo M, Okley G, Affum H, Acquah S, Gbadago J, Senu J, Botwe BO (2011) Water quality and level of some heavy metals in water and sediments of Kpeshie Lagoon, La-Accra, Ghana. Res J Environ Earth Sci 3:487-497.

Aloupi M, Angelidis M (2001) Geochemistry of natural and anthropogenic metals in the coastal sediments of the island of Lesvos, Aegean Sea. Environ Pollut 113:211-219.

Birch GF, Hutson P (2009) Use of sediment risk and ecological/conservation value for strategic management of estuarine environments: Sydney estuary, Australia. Environ Manage 44:836-850.

Botwe BO, Kelderman P, Nyarko E, Lens PNL (2017a) Assessment of DDT, HCH and PAH Contamination and Associated Ecotoxicological Risks in Surface Sediments of Coastal Tema Harbour (Ghana). Mar Pollut Bull 115: 480-488

Botwe BO, Schirone A, Delbono I, Barsanti M, Delfanti R, Kelderman P, Nyarko E, Lens PNL (2017b) Radioactivity concentrations and their radiological significance in sediments of the Tema Harbour (Greater Accra, Ghana). J Radiat Res Appl Sci 10:63-71.

Burton GA (2002) Sediment quality criteria in use around the world. Limnology 3:65-76.

Caille N, Tiffreau C, Leyval C, Morel JL (2003) Solubility of metals in an anoxic sediment during prolonged aeration. Sci Total Environ 301:239-250.

Calmano W (1988) Stabilization of dredged mud. Environmental Management of Solid Waste, Springer, Berlin, pp 80-98

Calmano W, Hong J, Foerstner U (1993) Binding and mobilization of heavy metals in contaminated sediments affected by pH and redox potential. Water Sci Technol 28:223-223.

Casado-Martínez MC., Buceta JL, Belzunce MJ, DelValls TA (2006) Using sediment quality guidelines for dredged material management in commercial ports from Spain. Environ Int 32:388-396.

Castro ÍB, Arroyo MF, Costa PG, Fillmann G (2012) Butyltin compounds and imposex levels in Ecuador. Arch Environ Contam Toxicol 62:68-77.

Chatterjee M, Silva Filho EV, Sarkar SK, Sella SM, Bhattacharya A, Satpathy KK, Prasad MVR, Chakraborty S, Bhattacharya BD (2007) Distribution and possible source of trace elements in the sediment cores of a tropical macrotidal estuary and their ecotoxicological significance. Environ Int 33:346-356.

De Jonge M, Teuchies J, Meire P, Blust R, Bervoets L (2012) The impact of increased oxygen conditions on metal-contaminated sediments part I: Effects on redox status, sediment geochemistry and metal bioavailability. Water Res 46:2205-2214.

Di Palma L, Mecozzi R (2007) Heavy metals mobilization from harbour sediments using EDTA and citric acid as chelating agents. Journal of hazardous materials 147:768-775.

Dung TTT, Cappuyns V, Swennen R, Phung NK (2013) From geochemical background determination to pollution assessment of heavy metals in sediments and soils. Rev Environ Sci Bio/Technol 12:335-353.

Hamzeh M, Ouddane B, Daye M, Halwani J (2014) Trace Metal Mobilization from Surficial Sediments of the Seine River Estuary. Water Air Soil Pollut 225:1878-1892.

Ho HH, Swennen R, Cappuyns V, Vassilieva E, Van Gerven T, Van Tran T (2012a) Potential release of selected trace elements (As, Cd, Cu, Mn, Pb and Zn) from sediments in Cam River-mouth (Vietnam) under influence of pH and oxidation. Sci Total Environ 435:487-498.

Ho HH, Swennen R, Cappuyns V, Vassilieva E, Van Tran T (2012b) Necessity of normalization to aluminum to assess the contamination by heavy metals and arsenic in sediments near Haiphong Harbor, Vietnam. J Asian Earth Sci 56:229-239.

Horowitz AJ (1985) A primer on trace metal-sediment chemistry, U.S. Geological Survey Water-Supply Paper 2277, 67 p

Iqbal J, Tirmizi S, Shah M (2013) Statistical apportionment and risk assessment of selected metals in sediments from Rawal Lake (Pakistan). Environ Monit Assess 185:729-743.

Jain C (2004) Metal fractionation study on bed sediments of River Yamuna, India. Water Res 38:569-578.

JICA (Japan International Co-operation Agency) Draft Environmental Impact Statement for Short-Term Development Plan of Tema Port (http://open_jicareport.jica.go.jp/pdf/11681632_03.pdf). Accessed 11 January 2017

Kelderman P, Osman A (2007) Effect of redox potential on heavy metal binding forms in polluted canal sediments in Delft (The Netherlands). Water Res 41:4251-4261.

Lepland A, Andersen TJ, Lepland A, Arp HPH, Alve E, Breedveld GD, Rindby A (2010) Sedimentation and chronology of heavy metal pollution in Oslo harbor, Norway. Mar Pollut Bull 60:1512-1522.

Long ER, Ingersoll CG, MacDonald DD (2006) Calculation and uses of mean sediment quality guideline quotients: a critical review. Environ Sci Technol 40:1726-1736.

Long ER, MacDonald DD, Smith SL, Calder FD (1995) Incidence of adverse biological effects within ranges of chemical concentrations in marine and estuarine sediments. Environ Manage 19: 81-97.

Mahu E, Nyarko E, Hulme S, Coale KH (2015) Distribution and enrichment of trace metals in marine sediments from the Eastern Equatorial Atlantic, off the Coast of Ghana in the Gulf of Guinea. Mar Pollut Bull 98:301-307.

Matijević S, Kušpilić G, Kljaković-Gašpić Z (2007) The redox potential of sediment from the Middle Adriatic region. Acta Adriatica, 48:191-204.

Mestres M, Sierra J, Mösso C, Sánchez-Arcilla A (2010) Sources of contamination and modelled pollutant trajectories in a Mediterranean harbour (Tarragona, Spain). Mar Pollut Bull 60:898-907.

Nyarko E, Botwe BO, Lamptey E, Nuotuo O, Foli BA, Addo MA (2011) Toxic metal concentrations in deep-sea sediments from the jubilee oil field and surrounding areas off the western coast of Ghana. Trop Environ Res 9:584-595.

Nyarko E, Fletcher A, Addo S, Foli BAK, Mahu E (2014) Geochemical Assessment of Heavy Metals in Surface Sediments: A Case Study of the Tema Port, Ghana. J Shipping Ocean Eng 4:79-92.

Phillips CR (2007) Sediment contaminant patterns within coastal areas of the Southern California Bight: multivariate analyses of Bight'98 Regional Monitoring Data. Bull Southern California Acad Sci 106:163-178.

Pini J, Richir J, Watson G (2015) Metal bioavailability and bioaccumulation in the polychaete Nereis (Alitta) virens (Sars): The effects of site-specific sediment characteristics. Mar Pollut Bull 95:565-575.

Pozza MR, Boyce JI, Morris WA (2004) Lake-based magnetic mapping of contaminated sediment distribution, Hamilton Harbour, Lake Ontario, Canada. J Appl Geophysics 57:23-41.

Pueyo M, Rauret G, Lück D, Yli-Halla M, Muntau H, Quevauviller P, Lopez-Sanchez J (2001) Certification of the extractable contents of Cd, Cr, Cu, Ni, Pb and Zn in a freshwater sediment following a collaboratively tested and optimised three-step sequential extraction procedure. J Environ Monit 3:243-250.

Qu W, Kelderman P (2001) Heavy metal contents in the Delft canal sediments and suspended solids of the River Rhine: multivariate analysis for source tracing. Chemosphere 45:919-925.

Quevauviller P, Rauret G, López-Sánchez JF, Rubio R, Ure A and Muntau H (1997) Certification of trace metal extractable contents in a sediment reference material (CRM 601) following a three-step sequential extraction procedure. Sci Total Environ 205:223-234.

Schipper C, Rietjens I, Burgess R, Murk A (2010) Application of bioassays in toxicological hazard, risk and impact assessments of dredged sediments. Mar Pollut Bull 60:2026-2042.

Schumacher BA (2002) Methods for the determination of total organic carbon (TOC) in soils and sediments. Ecological Risk Assessment Support Center, p1-23.

Taylor S (1964) Abundance of chemical elements in the continental crust: a new table. Geochim Cosmochim Acta 28:1273-1285.

Wang X-C, Zhang Y-X, Chen RF (2001) Distribution and partitioning of polycyclic aromatic hydrocarbons (PAHs) in different size fractions in sediments from Boston Harbor, United States. Mar Pollut Bull 42:1139-1149.

Chapter 6

Integrated Hazard, Risk and Impact Assessment of Tropical Marine Sediments from Tema Harbour (Ghana)

A modified version of this chapter was published as: Botwe, B.O., De Schamphelaere, K., Schipper, C.A., Teuchies, J., Blust, R., Nyarko, E., & Lens, P.N.L. (2017). Integrated hazard, risk and impact assessment of tropical marine sediments from Tema Harbour (Ghana). *Chemosphere, 177, 24-34*.

Abstract

The potential ecological hazard, risk and impact of tropical marine sediments from the Tema Harbour (Greater Accra, Ghana) was investigated by integrating *Corophium volutator* and *Hediste diversicolor* whole-sediment toxicity bioassays with data on the metals (Cd, Pb, Cr, Ni, Cu, Zn and As) concentrations of the sediments. The whole-sediment toxicity bioassay results showed that sediments of the Tema Harbour are potentially hazardous to marine benthic invertebrates. *C. volutator* exhibited a higher vulnerability to the sediment toxicity than *H. diversicolor*, although the latter showed higher biota-sediment accumulation factors for the investigated metals. Statistically significant correlations were observed between *C. volutator* mortality and sediment Cd concentration ($r = 0.84$, $p < 0.05$; $n = 6$) and between *H. diversicolor* mortality and sediment Cu concentration ($r = 0.94$, $p < 0.05$; $n = 5$). Comparison of metal concentrations with international action levels for contaminated sediment disposal indicates that the Tema Harbour sediments contain potentially hazardous concentrations of Cu and Zn. This study shows that sediments from the Tema Harbour are not suitable for disposal at sea without remediation. There is, therefore, a need to improve environmental management and regulate the disposal of dredged material originating from the Tema Harbour.

6.1.   Introduction

Contaminated sediments can be a source of hazardous contaminants to aquatic organisms, particularly benthic species (Burgess et al., 2007; Birch & Hutson, 2009). These benthic organisms play important roles in the functioning of aquatic ecosystems, such as biogeochemical cycling (Durou et al., 2007) and as a source of food for other species in the aquatic food chain (Burton Jr, 2002; Birch & Hutson, 2009; Carvalho et al., 2012; Gaion et al., 2014). The impact of contaminated sediments on benthic organisms can thus have serious consequences for the entire food chain (Burton Jr, 2002; Gaion et al., 2014) and the proper functioning of aquatic ecosystems. Consequently, sediment contamination is a major issue and information on the associated potential adverse ecological impact is of great interest to environmental regulators (Birch & Hutson, 2009; Schipper et al., 2010). Several biological effect-based sediment quality guidelines (SQGs) have been developed as predictive tools for assessing the potential of contaminated sediments to cause adverse biological effects (Burton Jr, 2002; Long et al., 2006; Schipper et al., 2013).

The abilities of SQGs to predict adverse biological effects associated with contaminated sediments are, however, limited since SQGs do not account for: (1) contaminant bioavailability (Schipper et al., 2010); (2) synergistic or antagonistic effects of contaminant mixtures present in sediments under natural conditions (Ciarelli et al., 1998; Forrester et al., 2003; Eisentraeger et al., 2004; Schipper et al., 2010); (3) multiple effects that may be exhibited by a single contaminant (Eggen et al., 2004); (4) chronic effects that may result from long-term exposure to low concentrations of contaminants in sediments (Eggen et al., 2004) and (5) contaminants present in sediments without being measured or identified as toxic or hazardous substances (Eisentraeger et al., 2004; Burgess et al., 2007). Consequently, whole-sediment toxicity bioassays have been recommended for the ecotoxicological characterisation of contaminated sediments to overcome the limitations of the SQG approach (Annicchiarico et al., 2007; Ré et al., 2009; Schipper et al., 2010). Whole-sediment toxicity bioassays involve the exposure of pollution-sensitive benthic invertebrates to contaminated sediments under laboratory conditions (Forrester et al., 2003). Integrating whole-sediment bioassays with the SQG approach can provide valuable insight into contaminants potentially contributing to sediment toxicity.

Marine benthic invertebrates such as *C. volutator* (Stronkhorst et al., 2003; Scarlett et al., 2007; van den Heuvel-Greve et al., 2007; Mayor et al., 2008) and *H. diversicolor* (Moreira et al., 2006; Mayor et al., 2008) are often employed as bio-indicators of pollution in marine and estuarine whole-sediment toxicity bioassays. Preference for *C. volutator* is due to its ease of collection and maintenance under laboratory conditions, availability in the field throughout the year, and tolerance to a wide range of salinities, sediment grain sizes and organic carbon contents (Ciarelli et al., 1998; Roddie & Thain, 2002; Scaps, 2002; Bat, 2005). *H. diversicolor* commonly occurs in intertidal areas, is able to survive in hypoxic and contaminated environments and exhibits tolerance to wide fluctuations in salinity and temperature (Scaps, 2002; Philippe et al., 2008). Both *C. volutator* and *H. diversicolor* have wide geographic distributions across polar, temperate and tropical marine regions (Bat, 2005; Moreira et al., 2006; Uwadiae, 2010; Carvalho et al., 2012). However, standard whole-sediment toxicity bioassay protocols have been developed mainly with temperate *C. volutator* (Roddie & Thain 2002; Schipper et al., 2006) and *H. diversicolor* (Hannewijk et al., 2004) with mortality as toxic response (endpoint), whereas whole-sediment toxicity bioassays with tropical species are not well developed (Adams and Stauber, 2008). Therefore, studies on the use of *C.*

volutator and *H. diversicolor* bioassays to assess the toxicity of sediments from tropical marine environments are scarce.

With over 50% of the world's population living in coastal zones (Petrosillo et al., 2009), the coastal marine environment is characterised by intense anthropogenic activities such as waste disposal, crude oil extraction and oil spills, shipping, fishing, agriculture and industrialisation (Petrosillo et al., 2009; Lepland et al., 2010; Schipper et al., 2010). This is also the case for the tropical marine Tema Harbour in Greater Accra (Ghana). Anthropogenic activities are a source of a wide range of hazardous substances, which adversely impact organisms inhabiting the coastal marine environment (Petrosillo et al., 2009; Lepland et al., 2010; Schipper et al., 2010): previous studies have shown that sediments of the Tema Harbour are contaminated by polycyclic aromatic hydrocarbons and organochlorine pesticide residues (Botwe et al., 2017) and metals (Nyarko et al., 2014; Botwe et al., unpublished results). Since Tema Harbour sediments are dredged periodically with subsequent disposal/storage under seawater, assessment of sediment quality is required to guide sediment management decisions at Tema Harbour and minimise adverse ecological impact. Therefore, the objectives of this study were to investigate: (1) the overall potential toxicity (hazard) of Tema Harbour sediments, (2) the potential risk (toxicity and bioavailability) of metal contamination in the sediments and (3) the potential impact (bioaccumulation) of metal contamination in the Tema Harbour sediments on benthic invertebrates by integrating whole-sediment toxicity bioassays with metal contamination data.

6.2. Materials and methods

6.2.1. Study area

The Tema Harbour in Greater Accra (Ghana) is a semi-enclosed coastal marine harbour with a water area of approximately 2 km^2, which forms part of the Gulf of Guinea (Fig. 6.1). The salinity of the Tema Harbour water ranges from 30 to 35 ‰. The Tema Harbour is compartmentalised into a Main Harbour, an Inner Fishing Harbour, an Outer Fishing Harbour and a Canoe Basin, which are bound to experience different anthropogenic impacts. The Main Harbour, the Inner Fishing Harbour and the Canoe Basin have been in operation since 1962, while the Outer Fishing Harbour was constructed in 1965. Various ships including oil tankers, bulk carriers, general cargo ships and containerships call at the Main Harbour. The Fishing Harbour provides handling facilities for semi-industrial and industrial fishing vessels such as

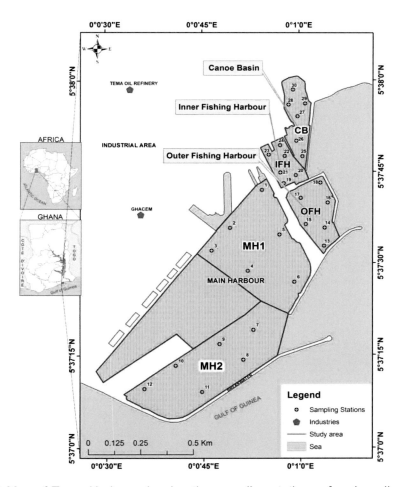

Fig. 6.1 Map of Tema Harbour showing the sampling stations of grab sediments (1-30). Grab sediments were composited into five samples for analysis as follows: sediment sample MH1 = grabs 1-6; MH2 = grabs 7-12; OFH = grabs 13-18; IFH = grabs 19-24 and CB = grabs 25-30.

trawlers, tuna vessels, and deep-sea carriers, while the Canoe Basin is a dedicated artisanal canoe fishing landing site. To ensure safe navigation, the Main Harbour is subject to dredging since 1998, whereas the Canoe Basin was dredged in May 2013. No dredging has yet been conducted in the Fishing Harbour. Located within an industrial setting, the Tema Harbour is subject to contamination not only from maritime operations (e.g. bunkering and refuelling, maintenance and repairs of vessels), but also from industrial activities (e.g. wastewater discharges into the harbour).

6.2.2. Sediment sampling

Grab sediment samples were collected from thirty stations (1-30) within the Tema Harbour (Fig. 6.1) in January 2016 using a stainless steel 3.5 L Ekman grab. The grab samples were composited to obtain five samples for analysis. In the Main Harbour, grabs 1-6 were composited to form sample MH1, while grabs 7-12 formed sample MH2. All grabs from the Outer Fishing Harbour (13-18) were composited into sample OFH, grabs from the Inner Fishing Harbour (19-24) formed sample IFH, while grabs from the Canoe Basin (25-30) formed sample CB (Fig. 6.1). All composite samples were mixed thoroughly with a plastic shovel in acid-washed plastic bowls before taking about 3.5 L portions into 3.78 L FoodSaver® zipper bags. All the samples were sealed air-tight using a hand-held vacuum pump and kept on ice in an ice-cool box in the field and during transportation to the Marine and Fisheries Department laboratory at the University of Ghana (Accra, Ghana), where they were stored overnight in a refrigerator at 4 °C. The samples were kept chilled in an ice-cool box and transported by air to the Systemic Physiological and Ecotoxicological Research laboratory (SPHERE) at the University of Antwerp (Belgium). The samples were kept there in a cold room at 4 °C until the bioassay experiments were conducted within 2 weeks of sample collection (Roddie & Thain, 2002).

6.2.3. Sampling of test organisms

The test organisms, *C. volutator* and *H. diversicolor*, were collected from the Eastern Scheldt located in the south-western part of the Netherlands, which is used as a non-contaminated control site for conducting whole-sediment bioassays (Kater et al., 2006; van den Heuvel-Greve et al., 2007). The field collection of *C. volutator* followed standard guidelines used around the world (Roddie & Thain, 2002). *C. volutator* were collected at low tide, when the shore was exposed, by carefully removing the upper 5 cm layer of sediments with a small stainless steel shovel into 10 L plastic buckets. The sediments were subsequently sieved over a 0.5 mm mesh with seawater from the same area (salinity of 30-31 ‰) into a separate 10 L plastic bucket while the *C. volutator* retained on the sieve were carefully transferred into another 10 L plastic bucket containing seawater. The sieved sediment (about 2 kg) was mixed thoroughly and kept as control sediment for the *C. volutator* bioassay.

H. diversicolor was carefully collected by hand along the banks of the Eastern Scheldt estuary, together with some of their associated sediments, and placed in a 10 L plastic bucket. The sediment was then covered by a layer of about 20 cm estuarine water. About 3 kg of sediment was also collected from the same area and sieved over a 0.5 mm mesh with some estuarine water from the same area into a 10 L plastic bucket to serve as control sediment for the *H. diversicolor* bioassay.

During sampling, care was taken to ensure that the *C. volutator* and *H. diversicolor* were not damaged. The collected samples were transported to the SPHERE laboratory, where the *C. volutator* and *H. diversicolor* aquaria were kept under continuous aeration at 15 °C in a climate-controlled room for 7 d for organisms to acclimatise prior to the whole-sediment bioassays. The water salinity in the aquaria was monitored with a conductivity meter (HACH, USA) during the acclimatisation period and kept at 30 ‰ by the addition of deionised water.

6.2.4. Laboratory bioassay experiments

The standard acute 10-day *C. volutator* whole-sediment bioassay as described by Roddie & Thain (2002) and Schipper et al. (2006) was adopted. The *C. volutator* used were of similar sizes (typically 4-5 mm long). The bioassays were conducted on sediments from the Tema Harbour and the reference site (control) in acid-washed 1.5 L wide-mouth glass bottles. Each set-up contained 200 mL homogenised sediment sample (about 3 cm thick), 600 mL well-aerated artificial seawater (about 12 cm depth of overlying seawater) of 30 ‰ salinity and 20 active individuals of *C. volutator*. Five (5) replicates were prepared for the control and the Tema Harbour sediment bioassays, except for the MH1 ($n = 4$) and OFH ($n = 3$) bioassays, due to loss of sample during transport.

The standard chronic 28-day *H. diversicolor* whole-sediment bioassay as described by Hannewijk et al. (2004) was adopted. *H. diversicolor* of similar mass (typically 0.2-0.3 g fresh weight) were used. The bioassays were conducted on sediments from the Tema Harbour and the reference site (control) in acid-washed 0.5 L wide-mouth glass bottles. Each set-up contained 100 mL homogenised sediment samples (about 3 cm thick), 120 mL artificially prepared seawater (about 6 cm depth of overlying seawater) and one active *H. diversicolor*. Fifteen (15) replicates were prepared for the control bioassay, while the number of bioassay replicates prepared per type of Tema Harbour sediment varied from 10 to 13, due to limited quantity of sediment sample. No CB sediment sample was available to conduct the *H. diversicolor* bioassay.

Prior to the introduction of *C. volutator* and *H. diversicolor*, the overlying seawater in each bottle was aerated continuously for 48 h to ensure adequate supply of oxygen, while avoiding sediment resuspension. Moreover, care was taken to ensure that the organisms were not damaged while being introduced into the bottles. All bioassays were conducted at a temperature of 15 ($\pm$ 1) °C and under a light regime of 16 h light and 8 h darkness. Periodic measurements of pH, salinity and dissolved oxygen (DO) levels of the overlying seawater were conducted from the start till the end of the exposure period using a pH meter, conductivity meter and DO meter (HACH, USA), respectively, and adjustments to the initial salinity were made when necessary. *H. diversicolor* were fed 3 times per week with 30 mg of ground fish food (TetraMin® XL Flakes) (Hannewijk et al., 2004), while their overlying seawater was renewed weekly to minimise the potential build-up of ammonia and hydrogen sulphide (Ferretti et al., 2000). At the end of their exposure periods, *C. volutator* were gently sieved over a 0.5 mm mesh, while *H. diversicolor* were gently removed with a pair of forceps. The organisms were then rinsed with artificial seawater to remove adhering particles and the numbers of living organisms were counted to determine mortalities.

6.2.5. Analyses of metal, sediment grain size and TOC contents

Upon completion of the bioassays, the organisms and sediment samples were freeze-dried prior to analyses of their metal contents. Metal analysis in *C. volutator* and *H. diversicolor* was based on whole-body tissues. *H. diversicolor* were analysed individually for metals. In the case of *C. volutator*, five composite samples were obtained from each replicate by pooling four individuals together and subsequently analysing their metal content. The freeze-dried *C. volutator* and *H. diversicolor* samples, together with 0.2 g portions of a mussel-based standard reference material (SRM 2976) from the National Institute of Standards and Technology (NIST, Luxembourg, Belgium) and procedural blanks were subjected to microwave-assisted digestion using 2 mL HNO$_3$ (for *H. diversicolor*) or 0.5 mL HNO$_3$ (for *C. volutator*).

About 0.2 g portions of freeze-dried homogenised sediment samples, a sediment certified reference material (BCR-701) and procedural blanks were analysed for total metal concentrations and metal fractionation by adopting the harmonised Community Bureau of

Reference (BCR) 3-step sequential extraction and *aqua regia* extraction techniques, respectively, following Pueyo et al. (2001). The metal concentrations were measured using ICP-MS (Varian, Australia). The dry sediments were also analysed for their grain-size distribution by the Malvern laser diffraction method (Blott et al., 2004), while total organic carbon (TOC) was analysed by the Walkey-Black wet oxidation method following Botwe et al. (2017). The mean metal recoveries in the standard BCR-701 samples ranged between 77 and 116 % with relative standard deviation (RSD) of 1.3-11.8 %, while the mean metal recoveries in the NIST SRM 2976 varied between 86 and 102% with RSD of 0.6-5.3 %, depending on the metal measured. Metal concentrations in the biota and sediment samples were corrected for recovery, using their respective mean recoveries in the certified reference materials.

6.2.6. Data analysis

One-way analysis of variance (ANOVA) and Holm-Sidak multiple comparison tests or Kruskal-Wallis one-way ANOVA on ranks (where normality test failed) were used to test for significant differences in *C. volutator* and *H. diversicolor* mortalities as well as metal concentrations in sediments across sites by using the statistical software SigmaPlot (version 11.0). Using the same software, normality and equal variance tests were performed with the Shapiro-Wilk and Levene's mean tests, respectively. Two-tailed Pearson correlations (using SPSS, version 16.0) among metal bioaccumulation factor, metal concentration in sediment, mortality and TOC content in sediment were determined separately for *C. volutator* ($n = 6$) and *H. diversicolor* ($n = 5$).

Regression plots were made to determine linear associations between measured parameters of interest using MS Excel 2007. To characterise the potential ecological impact of metal contamination in Tema Harbour sediments, biota-sediment accumulation factors (BSAFs) were estimated by dividing the metal concentrations in whole-body tissues of *C. volutator* and *H. diversicolor* by the corresponding concentrations in the sediment (Aydin-Onen et al., 2015). A BSAF >1.0 is indicative that metal bioaccumulation has occurred, and the greater the BSAF, the greater the bioaccumulation efficiency (Aydin-Onen et al., 2015). When necessary, raw metal concentrations were normalised to the <63 μm fraction of sediments as follows (Horowitz, 1985):

$$[\text{Metal}]n = \frac{100}{< 63 \ \mu\text{m fraction of sediment (\%)}} \times [\text{metal}]r \qquad (1)$$

where $[\text{Metal}]n$ is the normalised metal concentration and $[\text{metal}]r$ is the raw metal concentration.

6.3. Results

6.3.1. *C. volutator* and *H. diversicolor* mortalities in the whole-sediment bioassays

The mean *C. volutator* mortalities in the MH1, MH2, OFH, IFH, CB and control bioassays were 29, 38, 77, 99, 98 and 7 %, respectively (Fig. 6.2). One-way ANOVA ($p < 0.001$; $n = 6$) followed by the Holm-Sidak multiple comparison test ($p < 0.05$; $n = 6$) showed that the mortalities of *C. volutator* in the Tema Harbour sediment bioassays were significantly higher than those of the control bioassays with the following mortality trend across the Tema Harbour sediments: MH1 < MH2 < OFH < IFH ≈ CB. The *H. diversicolor* mortalities in the MH1, MH2, OFH and IFH bioassays were 46, 15, 10 and 92 %, respectively, while the control mortality was 7 % (Fig. 6.3). As in the case of *C. volutator*, the highest *H. diversicolor* mortality was associated with the IFH sediments. However, Fig. 6.3 shows that the trend in *H. diversicolor* mortality across the Tema Harbour sediments was OFH < MH2 < MH1 < IFH, which contrasted with the trend in *C. volutator* mortality (Fig. 6.2): the MH2 and OFH sediments caused low *H. diversicolor* mortalities (10-15 %) but high *C. volutator* mortalities (38-77 %).

6.3.2. Physicochemical conditions of water and sediments from the Tema Harbour and control bioassays

The physicochemical conditions of water and sediments from the Tema Harbour and control bioassays are summarised in Table 6.1. The salinity, pH and dissolved oxygen (DO) levels of the overlying water during the bioassay experiments varied between 28-34 ‰, 7.7-8.5 and 7.4-8.7 mgL^{-1}, respectively. There were marked differences in the grain size distributions of the Tema Harbour and control sediments: whereas sand (63-500 μm fraction) dominated the control sediment (62-90 %), a predominance (63-88 %) of silt (4-63 μm fraction) was present in the Tema Harbour sediments. However, the control sediments had a relatively higher TOC content (5.2-5.9 %) than the Tema Harbour sediments (2.9-4.2 %). Neither sediment grain

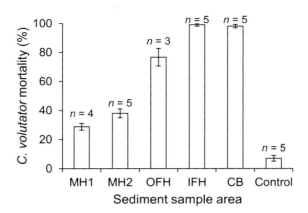

Fig. 6.2 Mean *C. volutator* mortalities in 10-day bioassay tests with whole-sediments from different areas within Tema Harbour, Ghana (MH1, MH2, OFH, IFH and CB) and the Eastern Scheldt, The Netherlands (control), indicating the number of replicates (*n*) and the standard error of the mean (error bars).

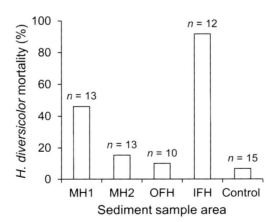

Fig. 6.3 *H. diversicolor* mortalities in 28-day bioassay tests with whole-sediments from different areas within Tema Harbour, Ghana (MH1, MH2, OFH and IFH) and the Eastern Scheldt estuary, The Netherlands (control), indicating the number of replicates (*n*).

size nor TOC content correlated significantly with both *C. volutator* mortality and *H. diversicolor* mortality.

Table 6.1 Physicochemical conditions of water and sediments in bioassays of bottom sediments from Tema Harbour, Ghana and Eastern Scheldt, The Netherlands (reference controls)

Sediment sample	Physicochemical conditions of water from bioassays			Sediment grain size distribution (%)			TOC (%)
	Salinity (‰)	pH	DO (mg/L)	<4 µm	4-63 µm	63-500 µm	
[a]MH1	30-32	8.0-8.4	7.9-8.4	10	79	11	3.2
[a]MH2	30-32	7.8-8.3	8.2-8.5	8	63	29	2.9
[a]OFH	30-33	7.7-8.4	7.9-8.1	12	85	3	3.6
[a]IFH	30-32	7.8-8.5	7.4-7.8	12	88	0	4.2
[a]CB	30-32	8.1-8.5	8.0-8.5	10	86	4	3.6
[b]RC	30-34	8.2-8.5	8.1-8.5	2	8	90	5.2
[c]RH	28-31	8.0-8.4	7.9-8.7	4	34	62	5.9

[a] sediments from Tema Harbour; [b] reference sediment from *C. volutator* site; [c] reference sediment from *H. diversicolor* site

6.3.3. Total metal concentrations and metal fractionation in the analysed sediments

The mean total metal concentrations in the Tema Harbour sediments (mg.kg^{-1} dw) ranged from 0.07 - 1.16 for Cd, 24.9-102 for Pb, 50.1-80.3 for Cr, 17.4-27.7 for Ni, 33.4-210 for Cu, 98-730 for Zn, and 7.9-14.2 for As (Table 6.2). One-way ANOVA ($p < 0.001$) followed by the Holm-Sidak multiple comparison test revealed significant differences ($p < 0.05$; $n = 6$) in the total concentrations of Pb and Zn (i.e., CB > IFH > OFH > MH1 > MH2), Cr (i.e., IFH > OFH > MH1 > CB > MH2) and Ni (i.e., OFH > IFH > MH1 > CB > MH2) across the Tema Harbour sediments. Kruskal-Wallis one-way ANOVA on ranks also revealed significant differences in the sediment concentrations of Cd ($p < 0.003$), Cu ($p < 0.004$) and As ($p < 0.004$) across the Tema Harbour bottom. The CB sediments had the highest Cd concentration, whereas the IFH sediments had the highest Cu and As concentrations. The mean concentrations for each metal for the Tema Harbour sediments were greater than the means for the Eastern Scheldt (control) sediments, except for Cd. A statistically significant correlation was observed between sediment Cd concentration and *C. volutator* mortality ($r =$

0.84, $p < 0.05$; $n = 6$), with a high coefficient of determination ($r^2 = 0.85$) (Fig. 6.4a). In the case of *H. diversicolor*, a statistically significant correlation was observed between mortality and sediment Cu concentration ($r = 0.94$, $p < 0.05$; $n = 5$), with a high coefficient of determination ($r^2 = 0.89$) (Fig. 6.4b).

Table 6.2 Mean ($n = 3$) concentrations of selected metals in sediments from Tema Harbour, Ghana and Eastern Scheldt, The Netherlands reference sites, and sediment quality guidelines (mg.kg^{-1} dw).

Sediment sample	Cd	Pb	Cr	Ni	Cu	Zn	As
[a]MH1	0.12 ± 0.00	39.7 ± 1.4	63.6 ± 1.6	24.2 ± 0.5	100 ± 0.4	190 ± 7	7.9 ± 0.2
[a]MH2	0.07 ± 0.00	24.9 ± 0.9	50.1 ± 0.8	17.4 ± 0.5	33.4 ± 0.5	98.0 ± 1.4	8.5 ± 0.2
[a]OFH	0.19 ± 0.00	50.6 ± 0.6	78.0 ± 0.2	27.7 ± 1.2	78.0 ± 2.1	244 ± 3	13.0 ± 0.2
[a]IFH	0.43 ± 0.03	84.3 ± 0.9	80.3 ± 0.3	26.0 ± 1.0	210 ± 7	415 ± 4	14.2 ± 0.5
[a]CB	1.16 ± 0.05	102 ± 2	61.2 ± 0.9	23.3 ± 0.5	195 ± 1	730 ± 5	12.6 ± 0.3
[a]RC	0.04 ± 0.01	4.6 ± 0.2	15.1 ± 0.5	3.0 ± 0.1	3.0 ± 0.2	14.7 ± 0.4	3.6 ± 0.1
[a]RH	0.21 ± 0.02	11.1 ± 0.7	20.4 ± 0.8	6.5 ± 0.6	6.5 ± 0.6	47.8 ± 4.4	5.7 ± 0.5
[b]ERL	1.2	46.7	81	20.9	34	150	8.2
[b]ERM	9.6	218	370	51.6	270	410	70

[a] same definition as in Table 6.1; [b] Sediment quality guidelines (DelValls et al., 2004); ERL represents a metal concentration associated with rare occurrence of harmful biological effects, the ERL-ERM interval represents a range of metal concentrations likely to cause harmful biological effects occasionally, while the ERM represents a metal concentration likely to cause harmful biological effects frequently (Long et al., 1995).

The fractionation of metals among exchangeable, reducible, oxidisable, and residual phases in the Tema Harbour sediments is shown in Table 6.3. Zn was present in appreciable amounts in all four fractions. The sediments contained considerable fractions of exchangeable metals only for Cd (15.7-46.8 %) and Zn (8.6-32.6 %). Cd (96.1-99.7 %), Pb (86.3-95 %), Cu (54.1-90.5 %) and Zn (72.2-96.5 %) were mainly present in the labile (i.e., the exchangeable, reducible and oxidisable) fractions rather than in the residual fraction. In contrast, Cr, Ni and As were predominantly present in the residual fraction, except for the IFH and CB sediments.

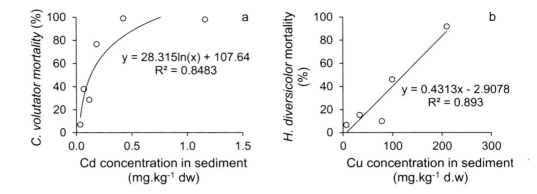

Fig. 6.4 Scatter plots of (a) *C. volutator* mortality versus Cd concentration in whole sediment (correlation coefficient, $r = 0.84$, $p < 0.05$; $n = 6$) and (b) *H. diversicolor* mortality versus Cu concentration in whole sediment (correlation coefficient, $r = 0.94$, $p < 0.05$; $n = 5$).

6.3.4. Metal bioaccumulation in *C. volutator* and *H. diversicolor*

Table 6.4 presents the mean concentrations of metals in whole-body tissues of *C. volutator* and *H. diversicolor* as well as their corresponding biota-sediment accumulation factors (BSAFs). The mean tissue Cd, Pb, Cr, Ni, Cu, Zn and As concentrations of *C. volutator* exposed to the Tema Harbour sediments were, respectively, 1-5, 4-17, 1-5, 2-5, 2-12, 2-7 and 1-5 times those of the controls. Similarly, the mean tissue Cd, Pb, Cr, Ni, Cu, Zn and As concentrations of *H. diversicolor* exposed to the Tema Harbour sediments were, respectively, 0-13, 4-10, 1-6, 1-3, 3-5, 3-4 and 2-4 times those of the controls. Generally, the metal BSAFs for *H. diversicolor* were markedly higher than those of C. volutator (except in some cases of Cd). In the tissues of C. volutator, Zn and Cu were found in the highest concentrations relative to the other metals, while Zn was found in the highest concentrations in *H. diversicolor*. However, for both *C. volutator* and *H. diversicolor*, the BSAFs of As were relatively higher than those of the other metals investigated. For all the metals investigated, no statistically significant correlations ($p > 0.05$) were observed between the sediment and tissue concentrations in *C. volutator* and *H. diversicolor*.

Table 6.3 Metal fractionation over the different phases in sediments from Tema Harbour, Ghana

Sediment sample	Sediment phase	Fraction of metals (%) associated with the different sediment phases						
		Cd	Pb	Cr	Ni	Cu	Zn	As
MH1	Exchangeable	46.8	2.8	0.3	4.0	1.5	20.8	2.8
	Reducible	30.7	79.6	18.7	8.6	42.8	48.4	20.6
	Oxidisable	19.6	6.3	21.3	16.1	30.0	11.4	9.9
	Residual	2.9	11.2	59.7	71.3	25.7	19.4	66.6
	Labile	97.1	88.8	40.3	28.7	74.3	80.6	33.4
MH2	Exchangeable	23.3	1.5	0.2	0.0	0.4	8.6	5.4
	Reducible	39.3	74.2	15.7	8.7	29.9	49.2	18.7
	Oxidisable	33.5	10.6	27.5	17.6	23.8	14.4	8.9
	Residual	3.9	13.7	56.6	73.7	45.9	27.8	66.9
	Labile	96.1	86.3	43.4	26.3	54.1	72.2	33.1
OFH	Exchangeable	36.0	1.2	0.2	4.3	1.0	21.7	1.6
	Reducible	46.8	78.8	20.7	12.3	41.4	53.4	25.5
	Oxidisable	14.7	7.5	24.4	18.9	32.7	10.8	13.8
	Residual	2.6	12.5	54.7	64.6	24.9	14.1	59.1
	Labile	97.4	87.5	45.3	35.4	75.1	85.9	40.9
IFH	Exchangeable	32.0	0.8	0.2	3.9	0.9	27.6	1.5
	Reducible	53.5	74.0	19.3	10.1	39.5	50.4	29.7
	Oxidisable	13.9	17.0	37.7	28.1	48.5	15.2	24.8
	Residual	0.6	8.2	42.9	57.8	11.2	6.8	43.9
	Labile	99.4	91.8	57.1	42.2	88.8	93.2	56.1
CB	Exchangeable	15.7	0.2	0.4	7.1	0.4	32.6	8.0
	Reducible	70.5	76.2	25.5	11.2	41.0	51.7	39.2
	Oxidisable	13.4	18.6	28.3	26.5	49.0	12.2	24.9
	Residual	0.3	5.0	45.8	55.2	9.5	3.5	28.0
	Labile	99.7	95.0	54.2	44.8	90.5	96.5	72.0

Risk assessment code (RAC) based on the percentage of sediment-associated metal in the exchangeable phase is as follows (Jain, 2004): no risk (<1 %), low risk (1-10 %), medium risk (11-30 %), high risk (31-50) and very high risk (>50 %). Labile fraction = sum of the fractions in the exchangeable, reducible and oxidisable phases.

Table 6.4 Metal concentrations (mean ± SE, mg.kg^{-1} dw) in whole-body tissues of *C. volutator* and *H. diversicolor* exposed to sediments from Tema Harbour, Ghana and estimated metal biota-sediment accumulation factors (BSAFs)

Sediment sample	Mean metal concentrations (BSAFs)						
	Cd	Pb	Cr	Ni	Cu	Zn	As
C. volutator							
[a]MH1	0.08 ± 0.01 (0.7)	2.1 ± 0.02 (0.05)	2.0 ± 0.3 (0.03)	1.5 ± 0.1 (0.06)	51 ± 3 (0.5)	52 ± 5 (0.3)	19 ± 1 (2.4)
[a]MH2	0.13 ± 0.01 (1.9)	4.7 ± 1.7 (0.2)	3.0 ± 0.5 (0.06)	1.6 ± 0.1 (0.1)	60 ± 3 (1.8)	57 ± 3 (0.6)	22 ± 2 (2.6)
[a]OFH	0.22 ± 0.01 (1.2)	9.8 ± 1.2 (0.2)	6.7 ± 0.9 (0.1)	3.7 ± 0.4 (0.1)	240 130 (3.1)	160 ± 30 (0.7)	45 ± 1 (3.5)
[a]IFH	0.05 ± 0.02 (0.12)	6.4 ± 1.2 (0.1)	3.8 ± 0.1 (0.05)	1.8 ± 0.1 (0.07)	50 ± 14 (0.2)	55 ± 9 (0.1)	12 ± 6 (0.9)
[a]CB	0.11 ± 0.03 (0.1)	8.6 ± 1.6 (0.1)	4.0 ± 0.7 (0.07)	2.1 ± 0.3 (0.1)	66 ± 11 (0.3)	80 ± 12 (0.1)	18 ± 2 (1.4)
[b]RC	0.05 ± 0.01 (1.3)	0.6 ± 0.1 (0.1)	1.4 ± 0.2 (0.1)	0.7 ± 0.1 (0.2)	20 ± 1 (6.7)	23 ± 2 (1.6)	10 ± 1 (2.8)
H. diversicolor							
[a]MH1	0.61 ± 0.50 (5)	110 ± 30 (3)	110 ± 48 (2)	100 ± 30 (4)	620 ± 130 (6)	7900 ± 1000 (42)	910 ± 70 (115)
[a]MH2	0.04 ± 0.03 (0.6)	180 ± 60 (75)	110 ± 60 (2)	80 ± 28 (4.6)	960 ± 240 (29)	9100 ± 1400 (93)	1140 ± 160 (134)
[a]OFH	2.6 ± 1.8 (14)	240 ± 130 (5)	90 ± 70 (1.2)	110 ± 50 (4)	1040 ± 270 (13)	10900 ± 1200 (45)	1440 ± 200 (110)
[a]IFH	ND	260 ± 150 (3)	24 ± 12 (0.3)	47 ± 16 (2)	1090 ± 330 (5)	10400 ± 2100 (25)	730 ± 230 (51)
[c]RH	0.20 ± 0.12 (1)	24 ± 6 (2)	18 ± 7 (1)	37 ± 8 (6)	200 ± 20 (31)	2800 ± 400 (58)	390 ± 50 (68)

[a] sediments from Tema Harbour; [b] reference sediment from *C. volutator* site; [c] reference sediment from *H. diversicolor* site; ND = not detected

6.4. Discussion

6.4.1. Hazard potential of Tema Harbour sediments

This study showed that the mean *C. volutator* mortalities in the Tema Harbour sediment bioassays exceeded the mean control mortality by >20% (Fig. 6.2), indicating that all the Tema Harbour sediments were toxic or hazardous to *C. volutator* (EPA, 1998). According to the International Council for the Exploration of the Sea based on *C. volutator* whole-sediment bioassays (ICES, 2008), sediment toxicity is classified as "elevated" or "high concern" if the *C. volutator* mortality exceeds that of the control by >30% and >60%, respectively. Thus, the MH2 sediments had elevated toxicity, whereas the toxicities of the OFH, IFH and CB sediments were of high concern. During the exposure period, no burrowing activity was observed in the IFH and CB sediments as most of the *C. volutator* avoided these sediments and kept swimming in the water column. This behaviour, which was not observed in the control and the remaining sediments, is an indication that *C. volutator* avoided the highly toxic sediments (Bat & Raffaelli, 1998; Bat, 2005).

Higher mortalities of *H. diversicolor* were observed in the MH1 (46%) and IFH (92%) sediment bioassays than in the control (7%), indicating that the MH1 and IFH sediments were also hazardous to *H. diversicolor*. However, the mortalities of *H. diversicolor* in the MH2 (15%) and the OFH (10%) were within the acceptable mortality of 10-15% (Thain & Bifield, 2002; ICES, 2008). As in the case of *C. volutator*, no burrowing activity of *H. diversicolor* was observed in the IFH sediments, possibly due to the high toxicity of the sediment (Bat & Raffaelli, 1998).

C. volutator and *H. diversicolor* exhibited strikingly different vulnerabilities towards the toxicities of the MH2 and OFH sediments, which underscores the importance of using a battery of species in whole-sediment toxicity bioassays (Bat & Raffaelli, 1998; DelValls et al., 2004; Eisentraeger et al., 2004; Annicchiarico et al., 2007; Schipper et al., 2010). Figs. 6.2 and 6.3 show that the MH2 and OFH sediments caused higher mortalities of *C. volutator* (38 and 77 %, respectively) than *H. diversicolor* (15 and 10 %, respectively). This observation supports previous findings that amphipods such as *C. volutator* are more sensitive to sediment toxicity than polychaetes such as *H. diversicolor* (Long et al., 2006). Moreover, sediments are often contaminated by a range of toxicants (Burton Jr, 2002; Forrester et al., 2003; Long et al., 2006), which may potentially impose different toxic responses on *C. volutator* and *H.*

diversicolor. For example, Bat & Raffaelli (1998) have shown that *C. volutator* and the marine polychaete *Arenicola* sp. have different sensitivities to Cu, Cd and Zn. They observed that lethal concentrations (LC_{50}) of Cu, Cd and Zn differed for *C. volutator* (37, 14 and 32 mg.kg^{-1} dw, respectively) and *Arenicola* sp. (20, 35 and 50 mg.kg^{-1} dw, respectively). This indicates that *C. volutator* is more sensitive to Cd and Zn, but less sensitive to Cu, than *Arenicola* sp..

The lower sensitivity of *C. volutator* to Cu compared to *H. diversicolor* has also been observed by Mayor et al. (2008), who reported LC_{50} values of 193 and 75 mg.kg^{-1} ww, respectively. The levels of Cd in the Tema Harbour sediments were much lower than the LC_{50} value of Cd for *C. volutator*,reported by Mayor et al. (2008). The levels of Cu (except for the MH2 sediments) and Zn in the Tema Harbour sediments exceeded their corresponding LC_{50} values reported for *C. volutator* (Bat & Raffaelli, 1998). This suggests that Cu and Zn may play a significant role in the toxicity of the Tema Harbour sediments. Bat & Raffaelli (1998) also observed that the mortality of *C. volutator* increased with increasing sediment Cu, Cd and Zn concentrations. In this study, statistically significant correlations were observed between *C. volutator* mortality and sediment Cd concentration and between *H. diversicolor* mortality and sediment Cu concentration. Scatter plots show that the former correlation is logarithmic with a high regression coefficient (R^2) of 0.85, whereas the latter correlation is linear with a high R^2 of 0.88 (Figs. 6.4a & 6b). The variation in sediment Cd and Cu concentrations could explain the variation in *C. volutator* and *H. diversicolor* mortalities, respectively. In the Main Harbour, for example, the Cu level in the MH1 sediment was higher than that of the MH2 sediment, the former resulting in a correspondingly higher *H.diversicolor* mortality. Although the level of Cu in the MH2 sediment was lower than that of the OFH sediment, the latter resulted in a lower *H. diversicolor* mortality. Clearly, other factors apart from the measured metals may play a role in the sediment toxicity to *H. diversicolor*. A potential source of the Cu contamination in the harbour sediments is the use of Cu-based antifouling paints on marine crafts to mitigate biofouling (Mukherjee et al., 2009).

Despite their markedly different vulnerabilities to the toxicities of the MH2 and OFH sediments, both *C. volutator* and *H. diversicolor* clearly distinguished the IFH sediments as being highly toxic or hazardous. The high toxicity of the CB sediments indicated by the associated high *C. volutator* mortality (98 %) could not be confirmed by using *H. diversicolor*, due to limited sediment quantity. The high toxicities of sediments from the IFH,

OFH and the CB are a clear indication that these areas are the most polluted within the Tema Harbour, which may be due to intense anthropogenic activities. The IFH and the OFH sustain a productive fishing industry through the provision of handling facilities for semi-industrial and industrial fishing vessels as well as a storage facility for petroleum products, while the CB is a dedicated artisanal canoe landing site. Daily, an average of 125 vessels operate from the IFH and the OFH, while the CB is normally overcrowded.

Fishing-related activities such as refuelling, painting of vessels and canoes, discharges of industrial wastewater and dumping of solid and liquid wastes by the fisherfolk are potential sources of chemical contamination in the Tema Harbour. Petroleum contamination was visible in the sediments from the IFH and the CB, possibly as a result of oil spills during refuelling of fishing vessels and fuel leakage in these sampling areas. Since their construction in 1962, the IFH and the OFH have not been dredged and there is a potential for pollutant accumulation in these areas over the years. The CB was dredged in May 2013 and hence, the high contamination of its sediments is an indication of high anthropogenic influences in this area. Although the Main Harbour is also subject to pollution from industrial wastewater discharges, ship traffic, oil spills during bunkering as well as dry dock activities such as scraping and painting of vessels (Mestres et al., 2010), recent maintenance dredging in March-April 2014 might have removed polluted sediments resulting in lower toxicities.

The control mortalities of both *C. volutator* and *H. diversicolor* in the whole-sediment toxicity bioassays (7%) were acceptably low, compared to the maximum acceptable mortality of 10-15% (Roddie & Thain, 2002; Casado-Martinez et al., 2007; Thain & Bifield, 2002; ICES, 2008). Moreover, the salinity, pH, dissolved oxygen and temperature of the overlying water during the bioassay experiments were all within acceptable ranges (Roddie & Thain, 2002; Thain & Bifield, 2002; Hannewijk et al., 2004; Schipper et al., 2006).

6.4.2. Potential ecological risks of metal contamination in Tema Harbour sediments

The potential ecological risks posed by metal (Cd, Pb, Cr, Ni, Cu, Zn and As) contamination in the Tema Harbour sediments were characterised by applying a pair of sediment quality guidelines, namely the effects-range low (ERL) and effects-range median (ERM). The ERL represents a pollutant concentration with a high probability to rarely cause harmful biological effects, the ERL-ERM interval represents a range of pollutant concentrations with a high probability to cause harmful biological effects occasionally, while the ERM represents a pollutant concentration with a high probability to cause harmful biological effects frequently (Long et al., 1995; Burton Jr, 2002; Long et al., 2006; DelValls et al., 2004; Birch & Hutson, 2009). Based on the ERL and ERM values (Table 6.2), two risk quotients, the ERL quotient (ERLQ) and the ERM quotient (ERMQ), were derived for each measured metal according to Eq. (2) and (3), respectively:

$$ERLQ = \frac{\text{Concentration of metal X (mg. kg}^{-1} \text{ dw})}{\text{ERL value of metal X (mg. kg}^{-1} \text{ dw})} \qquad (2)$$

$$ERMQ = \frac{\text{Concentration of metal X (mg. kg}^{-1} \text{ dw})}{\text{ERM value of metal X (mg. kg}^{-1} \text{ dw})} \qquad (3)$$

When the ERLQ exceeded 1.0, the corresponding ERMQ was derived instead in order to define three levels of potential risk associated with the measured metal concentration as follows: low risk for ERLQ $\leq$ 1.0, moderate risk for ERMQ < 1.0, and high risk for ERMQ $\geq$ 1.0.

The results showed that for all the Tema Harbour sediments analysed, at least one of the investigated metals may pose a moderate or high risk of harmful biological effects (Table 6.5). In particular, the levels of Zn in sediments from the IFH and the CB pose high potential ecotoxicological risks and are, therefore, of potential concern. This further supports the suggestion that Zn may play a significant role in the toxicity of the Tema Harbour sediments, and identifies the IFH and the CB as priority areas for management and remediation attention in the Tema Harbour. The levels of Ni and Cu (except for the MH2 sediment, where they may pose low potential risks), as well as As (except for the MH1 sediment, where the potential risk is low) pose moderate potential ecotoxicological risks. The levels of Cd and Cr present low potential ecotoxicological risks. The levels of Pb in sediments from the IFH, OFH and the CB were associated with moderate potential ecotoxicological risks, while those of the MH1 and the MH2 sediments were associated with low potential ecotoxicological risks.

Table 6.5 Derived risk quotients to assess risks posed by metal levels in sediments from Tema Harbour, Ghana (see Section 6.4.2 for derivation)

Sediment sample	Risk quotients						
	Cd	Pb	Cr	Ni	Cu	Zn	As
MH1	0.10	0.85	0.79	**0.47**	**0.37**	**0.47**	0.97
MH2	0.06	0.53	0.62	0.83	0.98	0.65	**0.12**
OFH	0.15	**0.23**	0.96	**0.54**	**0.29**	**0.60**	0.19
IFH	0.35	**0.39**	0.99	**0.50**	**0.73**	1.01*	0.20
CB	0.97	**0.47**	0.76	**0.45**	**0.72**	1.79	0.18

Unbold = effects-range low quotients (ERLQs): Bold = effects-range median quotients (ERMQs).

Metals fractionate over different phases in sediment: the exchangeable (carbonate-bound), reducible (iron/manganese oxide-bound), oxidisable (organic/sulphide-bound) and residual/refractory (silicate/mineral-bound) phases. Metals in the exchangeable phase are the most weakly bound and represent the potentially bioavailable fraction (Jain 2004; Igari et al., 2012; Iqbal et al., 2013). Based on the percentage of metal in the exchangeable phase, also known as the risk assessment code (RAC), the potential risks were characterised as follows (Jain, 2004): no risk (<1 %), low risk (1-10 %), medium risk (11-30 %), high risk (31-50) and very high risk (>50 %). Thus, the measured metal concentrations in the Tema Harbour sediments pose the following potential ecological risks: medium-high risk for Cd; low to high risk for Zn; low risk for As; no to low risk for Pb, Ni and Cu; and no risk for Cr (Table 6.3).

The relatively higher exchangeable fractions of Cd (15.7-46.8 %) and Zn (8.6-32.6 %) (Table 6.3) indicate that Cd and Zn had the highest potential of entering the food chain (Jain, 2004; Igari et al., 2012; Iqbal et al., 2013). The predominance of Cd, Pb, Cu and Zn in the labile phase, rather than the residual phase, suggests that these metals are derived predominantly from anthropogenic sources (Jain, 2004). Ni was predominantly in the residual phase, suggesting that it is mainly of natural origin. With regards to Cr and As, a predominance of the labile phase was observed in sediments from the IFH and the CB, suggesting greater anthropogenic influences in these areas of the Tema Harbour.

6.4.3. Potential ecological impact of sediment-associated metal in Tema Harbour

Benthic organisms can accumulate sediment-associated contaminants e.g. through direct ingestion of sediments (Burton Jr, 2002) and impact other organisms and humans via food chain transfer of the accumulated metals (Marsden & Rainbow, 2004). Table 6.4 shows that in most cases, the metal BSAFs for *H. diversicolor* exceeded 1.0, indicating a significant bioaccumulation of the investigated metals by *H. diversicolor* (Aydin-Onen et al., 2015). The metal BSAFs for *H. diversicolor* were higher than those for *C. volutator*, which may be attributed to the longer exposure periods for *H. diversicolor* (28 d) than *C. volutator* (10 d) and potential variability in metal bioaccumulation by different species (Marsden & Rainbow, 2004) due to e.g. potential differences in the ability to store or eliminate the metals (Adams et al., 2011). For both *C. volutator* and *H. diversicolor*, the BSAFs of As were higher than those of the other investigated metals, suggesting that among the investigated metals, As was either most efficiently taken up or least regulated by both *C. volutator* and *H. diversicolor*.

In most cases, *C. volutator* and *H. diversicolor* exposed to the Tema Harbour sediments had higher tissue concentrations of metals than the controls. Thus, the Tema Harbour sediments can be a significant source of metal bioaccumulation for benthic organisms with potential adverse impact on the aquatic food chain. Although the degree of contamination in harbours may be evident from the contamination patterns in the sediments and from biomarkers (de Boer et al., 2001; Schipper et al., 2009), this is not always evident (Schipper et al., 2009). Contrary to findings of other studies (Bat & Raffaelli, 1998), no statistically significant correlations were found between metal bioaccumulation (or metal bioavailability) and the corresponding metal concentrations in sediment or TOC content in this study. Moreover, no statistically significant correlations were found between metal bioaccumulation and mortality for both *C. volutator* and *H. diversicolor*. This lack of correlation is possibly due to the potential of the organisms to regulate metal uptake (Adams et al., 2011). Burrowing behaviour appeared to influence metal bioaccumulation as the non-burrowing organisms (those exposed to the IFH and CB bioassays) tend to have lower metal bioaccumulation factors. This is expected as burrowing results in increased exposure of organisms to contaminants in sediments (Bat, 2005).

6.4.4. Dredged material management implications for Tema Harbour

The practice of harbour dredging with subsequent disposal at sea poses a potential hazard to biota in the receiving site as dredged materials are often found to contain hazardous concentrations of chemical contaminants (Marsden & Rainbow, 2004; Schipper et al., 2010). To guide the management and disposal of dredged materials, countries such as the Netherlands and Spain have developed sediment quality guidelines or action levels, which represent potentially hazardous concentrations of chemical pollutants (DelValls et al., 2004; Casado-Martinez et al., 2006; Kelderman, 2012; Schipper et al., 2013).

Currently, no regulatory standards have been established for dredged materials in Ghana, despite routine maintenance dredging with subsequent disposal or storage at sea. Thus, in addition to the bioassay tests, the Spanish action levels (action levels 1 and 2) for dredged materials (DelValls et al., 2004; Casado-Martinez et al., 2006) were compared to the data for this study. These Spanish action levels are based on the probability of their associated chemical concentrations to cause adverse effects in marine biota (Casado-Martinez et al., 2006) and are presented in Table 6.6. Based on these action levels, dredged materials may be classified into three categories, which can then influence decisions about their management and disposal. Category I dredged materials have pollutant concentrations below action level 1 (AL1) and their disposal at sea is allowed, whereas category II dredged materials have pollutant concentrations between AL1 and action level 2 (AL2) and thus, would require further assessments to determine their suitability for disposal at sea. Category III dredged materials have pollutant concentrations above AL2 and, therefore, would require isolation or disposal in a confined area (DelValls et al., 2004). Since the AL1 and AL2 are based on the <63 µm fraction of sediments (DelValls et al., 2004), it was necessary to normalise the raw metal concentrations in the Tema Harbour sediments to the <63 µm fraction as described in section 6.2.6.

Table 6.6 shows that the levels of Cd, Pb, Cr, Ni, Cu, Zn and As in sediments from the MH2 and OFH areas were below their corresponding AL1 values. Thus, in relation to the investigated metals, dredged materials from the MH2 and OFH areas may fall under category I, which can be disposed at sea. The levels of Cu in sediments from the MH1 and IFH were between the AL1 and AL2 values, which are potentially hazardous. Similarly, the levels of Cd, Cu and Zn in sediments from the CB were between their corresponding AL1 and AL2 values and are, therefore, potentially hazardous and thus, the disposal of these sediments at sea is inappropriate.

Table 6.6 Comparison of metal concentrations in Tema Harbour (Ghana) sediments (normalised to the <63 μm fraction) with Spanish sediment quality action levels (AL1 and AL2)

	Cd	Pb	Cr	Ni	Cu	Zn	As
[a]MH1	0.1	45	72	27	**112**	216	9
[a]MH2	0.1	35	71	25	47	138	12
[a]OFH	0.2	52	80	29	81	252	13
[a]IFH	0.4	85	81	26	**197**	416	14
[a]CB	**1.2**	106	64	24	**204**	**768**	13
[b]AL1	1	120	200	100	100	500	80
[b]AL2	5	600	1000	400	400	3000	200

[a] Tema Harbour sediments; [b] action levels (DelValls et al., 2004); unbold = [metal] < AL1 and indicates permissible metal levels for sediment disposal at sea; bold = AL1 < [metal] < AL2 and indicates sediments would require further assessment prior to sediment disposal at sea.

6.5. Conclusion

The standard 10-day *C. volutator* and 28-day *H. diversicolor* whole-sediment bioassays were combined with metal (Cd, Pb, Cr, Ni, Cu, Zn and As) concentrations to investigate the potential ecological hazard, risk and impact of contaminated sediments from the Tema Harbour. *C. volutator* was found more vulnerable than *H. diversicolor* to the toxicity effects the Tema Harbour sediments and underscores the importance of using different of species in whole-sediment toxicity bioassays. The concentrations of Cu and Zn may play a role in the mortalities of *C. volutator* and *H. diversicolor*. A logarithmic correlation was observed between sediment Cd concentration and *C. volutator* mortality, while a linear correlation was observed between sediment Cu concentration and *H. diversicolor* mortality. Risk assessment based on sediment quality guidelines indicated that the metal contamination in the Tema Harbour sediments poses moderate to high potential ecological risks. The results indicate a need to improve environmental management and regulate the disposal of dredged materials at the Tema Harbour.

References

Adams, M.S., & Stauber, J.L., 2008. Marine whole sediment toxicity tests for use in temperate and tropical Australian environments: current status. Aust. J. Ecotoxicol. 14, 155-167.

Adams, W.J., Blust, R., Borgmann, U., Brix, K.V., DeForest, D.K., Green, A.S., Meyer, J.S., McGeer, J.C., Paquin, P.R., & Rainbow, P.S., 2011. Utility of tissue residues for predicting effects of metals on aquatic organisms. Integr. Environ. Assess. Manage. 7, 75-98.

Annicchiarico, C., Biandolino, F., Cardellicchio, N., Di Leo, A., Giandomenico, S., & Prato, E., 2007. Predicting toxicity in marine sediment in Taranto Gulf (Ionian Sea, Southern Italy) using Sediment Quality Guidelines and a battery bioassay. Ecotoxicol. 16, 239-246.

Aydin-Onen, S., Kucuksezgin, F., Kocak, F., & Açik, S., 2015. Assessment of heavy metal contamination in *Hediste diversicolor* (OF Müller, 1776), Mugil cephalus (Linnaeus, 1758), and surface sediments of Bafa Lake (Eastern Aegean). Environ. Sci. Pollut. Res. 22, 8702-8718.

Bat, L., 2005. A review of sediment toxicity bioassays using the amphipods and polychaetes. Turkish J. Fish. Aquat. Sci. 5, 119-139.

Bat, L., & Raffaelli, D. (1998). Sediment toxicity testing: a bioassay approach using the amphipod *Corophium volutator* and the polychaete *Arenicola marina*. J. Exp. Mar. Bio. Ecol. 226, 217-239.

Birch, G.F., & Hutson, P., 2009. Use of sediment risk and ecological/conservation value for strategic management of estuarine environments: Sydney estuary, Australia. Environ. Manage. 44, 836-850.

Blott, S.J., Croft, D.J., Pye, K., Saye, S.E., & Wilson, H.E., 2004. Particle size analysis by laser diffraction. Geol. Soc. Spec. Publ. 232, 63-73.

Botwe, B. O., Kelderman, P., Nyarko, E., & Lens, P.N.L., 2017. Assessment of DDT, HCH and PAH contamination and associated ecotoxicological risks in surface sediments of coastal Tema Harbour (Ghana). Mar. Pollut. Bull. 115, 480-488.

Burgess, R.M., Perron, M.M., Cantwell, M.G., Ho, K.T., Pelletier, M.C., Serbst, J.R., & Ryba, S.A., 2007. Marine sediment toxicity identification evaluation methods for the anionic metals arsenic and chromium. Environ. Toxicol. Chem. 26, 61-67.

Burton Jr, G.A., 2002. Sediment quality criteria in use around the world. Limnology, 3, 65-76.

Carvalho, S., Cunha, M.R., Pereira, F., Pousão-Ferreira, P., Santos, M., & Gaspar, M., 2012. The effect of depth and sediment type on the spatial distribution of shallow soft-bottom amphipods along the southern Portuguese coast. Helgol. Mar. Res. 66, 489-501.

Casado-Martínez, M.C., Buceta, J.L., Belzunce, M.J., & DelValls, T.A., 2006. Using sediment quality guidelines for dredged material management in commercial ports from Spain. Environ. Int. 32, 388-396.

Casado-Martinez, M.C., Forja, J.M., & DelValls, T.A., 2007. Direct comparison of amphipod sensitivities to dredged sediments from Spanish ports. Chemosphere, 68, 677-685.

Ciarelli, S., Vonck, W., Van Straalen, N., & Stronkhorst, J., 1998. Ecotoxicity assessment of contaminated dredged material with the marine amphipod *Corophium volutator*. Arch. Environ. Contam. Toxicol. 34, 350-356.

de Boer, J., van der Zande, T. E., Pieters, H., Ariese, F., Schipper, C. A., van Brummelen, T., & Vethaak, A. D. (2001). Organic contaminants and trace metals in flounder liver and sediment from the Amsterdam and Rotterdam harbours and off the Dutch coast. J. Environ. Monit. 3, 386-393.

DelValls, T., Andres, A., Belzunce, M., Buceta, J., Casado-Martinez, M., Castro, R., Riba, I., Viguri, J.R., & Blasco, J., 2004. Chemical and ecotoxicological guidelines for managing disposal of dredged material. Trends Anal. Chem. 23, 819-828.

Durou, C., Poirier, L., Amiard, J.-C., Budzinski, H., Gnassia-Barelli, M., Lemenach, K., Peluhet, L., Mouneyrac, C., Roméo, M., & Amiard-Triquet, C., 2007. Biomonitoring in a clean and a multi-contaminated estuary based on biomarkers and chemical analyses in the endobenthic worm *Nereis diversicolor*. Environ. Pollut. 148, 445-458.

Eggen, R.I., Behra, R., Burkhardt-Holm, P., Escher, B.I., & Schweigert, N., 2004. Challenges in ecotoxicology. Environ. Sci. Technol. 38, 58A-64A.

Eisentraeger, A., Hund-Rinke, K., & Roembke, J., 2004. Proposal of a testing strategy and assessment criteria for the ecotoxicological assessment of soil or soil materials. J. Soils Sediments, 4, 123-128.

EPA, USACE, 1998. Evaluation of dredged material proposed for discharge in waters of the US-testing manual: Inland Testing Manual. EPA-823-B-98-004. US Environmental Protection Agency and US Army Corps of Engineers, Washington, DC.

Ferretti, J.A., Calesso, D.F., & Hermon, T.R., 2000. Evaluation of methods to remove ammonia interference in marine sediment toxicity tests. Environ. Toxicol. Chem. 19, 1935-1941.

Forrester, G.E., Fredericks, B.I., Gerdeman, D., Evans, B., Steele, M.A., Zayed, K., Schweitzer, L.E., Suffet, I.H., Vance, R.R., & Ambrose, R.F., 2003. Growth of estuarine fish is associated with the combined concentration of sediment contaminants and shows no adaptation or acclimation to past conditions. Mar. Environ. Res. 56, 423-442.

Gaion, A., Sartori, D., Scuderi, A., & Fattorini, D., 2014. Bioaccumulation and biotransformation of arsenic compounds in *Hediste diversicolor* (Muller 1776) after exposure to spiked sediments. Environ. Sci. Pollut. Res. 21, 5952-5959.

Hannewijk, A., Kater, B.J., Schipper, C.A., 2004. Sediment toxiciteitstest met *Nereis diversicolor* (In Dutch). 22 pp.

Horowitz, A. J., 1985. A primer on trace metal-sediment chemistry, U.S. Geological Survey Water-Supply Paper 2277, 67 p.

ICES, 2008. Report of the Fourth ICES/OSPAR Workshop on Integrated Monitoring of Contaminants and their Effects in Coastal and Open Sea Areas (WKIMON IV). ICES Document CM 2008/ACOM: 49. 71 pp.

Igari, Y., Ohno, M., Tamura, T., Suzuki, K., Kose, T., & Kawata, K., 2012. Chemical speciation of metals in surface sediments from small urban and agricultural rivers. Bull. Environ. Contam. Toxicol. 89, 764-769.

Iqbal, J., Tirmizi, S., & Shah, M., 2013. Statistical apportionment and risk assessment of selected metals in sediments from Rawal Lake (Pakistan). Environ. Monit. Assess. 185, 729-743.

Jain, C., 2004. Metal fractionation study on bed sediments of River Yamuna, India. Water Res. 38, 569-578.

Kater, B.J., Dubbeldam, M., & Postma, J.F., 2006. Ammonium toxicity at high pH in a marine bioassay using *Corophium volutator*. Arch. Environ. Contam. Toxicol. 51, 347-351.

Kelderman, P., 2012. Sediment pollution, transport, and abatement measures in the city canals of delft, The Netherlands. Water Air Soil Pollut. 223, 4627-4645.

Lepland, A., Andersen, T.J., Lepland, A., Arp, H.P.H., Alve, E., Breedveld, G.D., & Rindby, A., 2010. Sedimentation and chronology of heavy metal pollution in Oslo harbor, Norway. Mar. Pollut. Bull. 60, 1512-1522.

Long, E., Macdonald, D., Smith, S., & Calder, F., 1995. Incidence of adverse biological effects within ranges of chemical concentrations in marine and estuarine sediments. Environ. Manage. 19, 81-97.

Long, E.R., Ingersoll, C.G., & MacDonald, D.D., 2006. Calculation and uses of mean sediment quality guideline quotients: a critical review. Environ. Sci. Technol. 40, 1726-1736.

Marsden, I., & Rainbow, P., 2004. Does the accumulation of trace metals in crustaceans affect their ecology - the amphipod example? J. Exp. Mar. Bio. Ecol. 300, 373-408.

Mayor, D.J., Solan, M., Martinez, I., Murray, L., McMillan, H., Paton, G.I., & Killham, K., 2008. Acute toxicity of some treatments commonly used by the salmonid aquaculture industry to *Corophium volutator* and *Hediste diversicolor*: Whole sediment bioassay tests. Aquaculture, 285, 102-108.

Mestres, M., Sierra, J., Mösso, C., & Sánchez-Arcilla, A., 2010. Sources of contamination and modelled pollutant trajectories in a Mediterranean harbour (Tarragona, Spain). Mar. Pollut. Bull. 60, 898-907.

Moreira, S.M., Lima, I., Ribeiro, R., & Guilhermino, L., 2006. Effects of estuarine sediment contamination on feeding and on key physiological functions of the polychaete *Hediste diversicolor*: laboratory and in situ assays. Aquat Toxicol. 78, 186-201.

Mukherjee, A., Rao, K.M., & Ramesh, U. (2009). Predicted concentrations of biocides from antifouling paints in Visakhapatnam Harbour. J. Environ. Manag. 90, S51-S59.

Nyarko, E., Fletcher, A., Addo, S., Foli, B.A.K., & Mahu, E., 2014. Geochemical assessment of heavy metals in surface sediments: A case study of the Tema Port, Ghana. J. Ship Ocean Eng. 4, 79-92.

Petrosillo, I., Valente, D., Zaccarelli, N., & Zurlini, G., 2009. Managing tourist harbors: Are managers aware of the real environmental risks? Mar. Pollut. Bull. 58, 1454-1461.

Philippe, S., Leterme, C., Lesourd, S., Courcot, L., Haack, U., & Caillaud, J., 2008. Bioavailability of sediment-borne lead for ragworms (*Hediste diversicolor*) investigated by lead isotopes. Appl. Geochem. 23, 2932-2944.

Pueyo, M., Rauret, G., Lück, D., Yli-Halla, M., Muntau, H., Quevauviller, P., & Lopez-Sanchez, J., 2001. Certification of the extractable contents of Cd, Cr, Cu, Ni, Pb and Zn in a freshwater sediment following a collaboratively tested and optimised three-step sequential extraction procedure. J. Environ. Monit. 3, 243-250.

Ré, A., Freitas, R., Sampaio, L., Rodrigues, A.M., & Quintino, V., 2009. Estuarine sediment acute toxicity testing with the European amphipod Corophium multisetosum Stock, 1952. Chemosphere, 76, 1323-1333.

Roddie, B., & Thain, J., 2002. Biological effects of contaminants: *Corophium sp.* sediment bioassay and toxicity test: International Council for the Exploration of the Sea. ICES

Techniques in Marine Environmental Sciences No. 28, ISSN 0903-2606, Copenhagen, Denmark.

Scaps, P., 2002. A review of the biology, ecology and potential use of the common ragworm *Hediste diversicolor* (OF Müller)(Annelida: Polychaeta). Hydrobiologia, 470, 203-218.

Scarlett, A., Rowland, S., Canty, M., Smith, E., & Galloway, T., 2007. Method for assessing the chronic toxicity of marine and estuarine sediment-associated contaminants using the amphipod *Corophium volutator*. Mar. Environ. Res. 63, 457-470.

Schipper, C.A., Burgess, R.M., van den Dikkenberg, L.C., 2006. Sediment toxiciteittest met de slijkgarnaal *Corophium volutator* (In Dutch). 18 pp.

Schipper, C. A., Lahr, J., Van den Brink, P. J., George, S. G., Hansen, P.-D., de Assis, H. C. d. S., Van der Oost, R., Thain, J.E., Livingstone, D., & Mitchelmore, C., 2009. A retrospective analysis to explore the applicability of fish biomarkers and sediment bioassays along contaminated salinity transects. ICES J. Mar. Sci. 66, 2089-2105

Schipper, C., Rietjens, I., Burgess, R., & Murk, A., 2010. Application of bioassays in toxicological hazard, risk and impact assessments of dredged sediments. Mar. Pollut. Bull. 60, 2026-2042.

Schipper, C. A., Leonards, P. E. G., Klamer, H. J., Thomas, K. V., & Vethaak, A. D., 2013. Protocol for measuring dioxin-like activity in environmental samples using *in vitro* reporter gene DR-Luc assays. ICES Techniques in Marine Environmental Sciences No. 55. 21 pp.

Stronkhorst, J., Schipper, C., Brils, J., Dubbeldam, M., Postma, J., & van de Hoeven, N. (2003). Using marine bioassays to classify the toxicity of Dutch harbor sediments. Environ. Toxicol. Chem. 22, 1535-1547.

Thain, J., & Bifield, S., 2002. Biological effects of contaminants: sediment bioassay using the polychaete *Arenicola marina*: International Council for the Exploration of the Sea.

Uwadiae, R.E. (2010). An inventory of the benthic macrofauna of Epe Lagoon, southwest Nigeria. J. Sci. Res. Develop. 12, 161-171.

van den Heuvel-Greve, M., Postma, J., Jol, J., Kooman, H., Dubbeldam, M., Schipper, C., & Kater, B., 2007. A chronic bioassay with the estuarine amphipod *Corophium volutator*: Test method description and confounding factors. Chemosphere, 66, 1301-1309.

Chapter 7

Settling Fluxes and Sediment Accumulation Rates by the Combined Use of Sediment Traps and Sediment Cores in Tema Harbour (Ghana)

A modified version of this chapter was published as: Botwe, B.O., Abril, J.M., Schirone, A., Barsanti, M., Delbono, I., Delfanti, R., Nyarko, E., & Lens, P.N.L. (2017). Settling fluxes and sediment accumulation rates by the combined use of sediment traps and sediment cores in Tema Harbour (Ghana). *Science of the Total Environment, 609*, 1114-1125.

Abstract

Settling fluxes and sediment accumulation rates in coastal Tema Harbour (Ghana) were investigated by the combined analyses of results in sediment traps and sediment cores. Sediment traps were deployed at 5 stations within the Tema Harbour at 2 sampling depths and were retrieved every two weeks till the end of 12 weeks to estimate the Settling Fluxes (SFs). Four sediment cores from the harbour were analysed for their radioactivity (^{7}Be, ^{234}Th, ^{210}Pb, ^{212}Pb, ^{226}Ra, ^{40}K and ^{137}Cs) profiles to quantify Sediment Accumulation Rates (SARs). The sediment cores exhibited variable bulk density profiles, indicating highly dynamic and non-steady sedimentation conditions. ^{7}Be-derived gross-estimates of very recent SARs using the constant flux-constant sedimentation (CF-CS) model were in the range of 2.5-9.0 g.cm^{-2}.y^{-1}. These values were much lower than the estimated average SFs (15.2-53.8 g.cm^{-2}.y^{-1}), indicating sediment resuspension plays an important role. On a decadal time scale, conventional ^{210}Pb sediment dating models did not allow any estimation of SARs in the Tema Harbour. Thus, the ^{210}Pb-based TERESA model was applied to depict a reliable scenario for sedimentation with time-averaged SARs in the range of 1.4-3.0 g.cm^{-2}.y^{-1} and fluxes of matter contributed by the marine inflow and local sources. Sediment accretion rates of 1.7-3 cm.y^{-1} were also inferred, which may pose a moderate problem of sustainability for the Tema Harbour. This study reveals how the geochemical behaviour of different radionuclides with Gamma spectrometry in the marine environment can be used to obtain reliable information on the complex dynamics of Suspended Particulate Matter (SPM), even in a very disturbed and anthropic environment as a coastal harbour area where (1) conventional ^{210}Pb-based dating methods fail and (2) the use of sediment traps and ^{234}Th and ^{7}Be profiles in sediment cores show serious constraints.

7.1. Introduction

Coastal marine harbours support national economies through shipping, fishing and tourism and are, therefore, considered important assets for coastal nations (Van Rijn, 2005). Coastal marine harbours are prone to large influxes of sediments as a result of waves and tidal currents (Leys and Mulligan, 2011), as well as erosion downdrift of the harbour breakwater. In semi-enclosed coastal harbours, the restricted water movement may enhance the settling of sediments and result in their accumulation within the harbour basin (Lepland et al., 2010; Botwe et al., 2017a). Sediment accumulation in harbours poses navigation and ecological challenges (Syvitski et al., 2005; Van Rijn 2005; Green and Coco, 2014), requiring dredging at high costs (Qu and Kelderman, 2001; Barneveld and Hugtenburg, 2008; Schipper et al., 2010). Human activities around harbours often result in the accumulation of a variety of hazardous pollutants in sediments (Botwe et al. 2017a, 2017b), which poses serious concerns about the handling and fate of the dredged materials. Sedimentation in harbours is thus a major issue and its quantification is essential for harbour management (Buesseler et al., 2007; Leys and Mulligan, 2011).

Information on sediment accumulation rates (SARs) in harbours, reservoirs, estuaries and coastal areas can be obtained from the comparison of bathymetric data acquired at different periods (Khaba and Griffiths, 2017, Brucker et al., 2007). This GIS-bathymetric approach involves measurements of tidal levels, water depth, positioning and the application of a series of corrections. For shallow waters with less than 20 m water depth such as harbours, bathymetric data can be best obtained by using Phase Differencing Bathymetric Sonar (PDBS) systems with an associated total vertical uncertainty of 0.26 m for a 10 m water depth (Brisson et al., 2014). Although this uncertainty is acceptable for water depth control purposes, it is too coarse for estimating SARs. Thus, GIS-bathymetry has been mainly used for mapping SAR in areas where accretion rates are high (>0.3 m.y^{-1}) (Brucker et al., 2007) or when the study involves time lapses of the order of decades (Ortt et al., 2000). It is worth noting that the method provides mean SAR in the time lapse and it cannot identify processes affecting the depth distribution of hazardous pollutants. Other methods based on horizon markers, anchored tiles, rulers, sediment traps, optical backscatter sensors and short-lived radionuclides (^{234}Th and ^{7}Be) are available for measuring sediment accumulation over short-time scales (Thomas and Ridd, 2014).

In depositional environments, where continuous accumulation of sediments has taken place over a long period of time without any interruption in the sedimentary sequence or mixing, the ^{210}Pb sediment dating technique has proven useful for quantifying SARs on time scales spanning 100-150 years (Appleby and Oldfield, 1978; Robbins, 1978; Appleby and Oldfield, 1983; Caroll and Lerche, 2003; Alonso-Hernandez et al., 2006; Díaz-Asencio et al., 2011; Bellucci et al., 2012). Sedimentation in coastal marine harbours is, however, a dynamic process for which relatively large spatial and temporal variations can be expected (Leys and Mulligan, 2011). For example, sediment resuspension due to wave action, tidal oscillations, movement of vessels and dredging (Lepland et al., 2010; Leys and Mulligan, 2011; Green and Coco, 2014) may co-occur with sediment deposition in coastal marine harbours. Thus, complete recovery of radionuclide inventories could also be difficult in areas with relatively high SARs (of the order of 1 $g.cm^{-2}.y^{-1}$) due to the limited length of sediment core samplers. Quantifying sedimentation rates in such disturbed environments, therefore, requires integrated approaches (Bellucci et al., 2012), the application of robust numerical models and an accurate study of the harbour and the operations within the harbour itself (Tanner et al., 2000; Tang et al., 2008; Smith et al., 2009; Lepland et al., 2010).

As far as we know, there are no previous studies on harbours located in tropical Africa, where in addition to the aforementioned challenges and the scarcity of background studies, it is difficult to measure artificial radionuclides such as ^{137}Cs above the method detection limit due to its low fallout rate. ^{210}Pb-based chronologies require independent chronostratigraphic markers for validation (Smith, 2001; Caroll and Lerche, 2003; Abril, 2004) and ^{137}Cs is the most widely used complement for the ^{210}Pb dating method.

Profiles of short-lived and particle-reactive radionuclides such as ^{234}Th (half-life = 24 d) and ^{7}Be (half-life = 53 d) in sediment cores can be used to estimate recent sedimentation rates over short-time scales of days to months (Giffin and Corbett, 2003). ^{234}Th is produced from the decay of ^{238}U (half-life = 4.5 x 10^9 y), while ^{7}Be is a cosmogenic radionuclide produced from the interaction of cosmic rays with oxygen and nitrogen in the stratosphere and the troposphere (Sharma et al., 1987; Erten, 1997; Pfitzner et al., 2004; Palinkas et al., 2005). In marine environments, ^{238}U is found in the water column in excess with respect to ^{226}Ra due to its higher solubility (IAEA, 2004; Botwe et al., 2007b). When ^{238}U decays to ^{234}Th, this high particle-reactive isotope is taken up by Suspended Particulate Matter (SPM) and surface sediments, where it can be found in excess with respect to the background levels, referred to as excess ^{234}Th ($^{234}Th_{exc}$). ^{7}Be and $^{234}Th_{exc}$ profiles, when combined with ^{210}Pb and ^{137}Cs profiles in sediment cores, are useful for understanding sediment dynamics in aquatic systems (Sharma et al., 1987; Erten, 1997; Fuller et al., 1999; Giffin and Corbett, 2003; Palinkas et al., 2005; Yeager et al., 2005; Schmidt et al., 2007a; Schmidt et al., 2007b). In addition to radionuclide profiles in sediment cores, sediment traps are important tools for investigating short-term fluxes and dynamics of sediments in aquatic systems (Kelderman, 2012; Kelderman et al., 2012; de Vicente et al., 2010).

The objective of this study was to investigate settling fluxes and sedimentation rates in the coastal marine Tema Harbour (Ghana) by the combined analyses of accumulated dry mass in sediment traps and radionuclide ([7]Be, [234]Th, ^{210}Pb, [212]Pb, [226]Ra, [40]K and ^{137}Cs) profiles in sediment cores, with the help of numerical sediment dating models. In particular, excess [210]Pb ([210]Pb$_{exc}$) data were analysed with the TERESA (Time Estimates from Random Entries of Sediments and Activities) model (Abril, 2016), which is based on a widely observed statistical correlation between [210]Pb$_{exc}$ fluxes and SARs (Abril and Brunskill, 2014). The TERESA model has been validated against synthetic cores and real data from varved sediments, for which an independent chronology was available. The basic assumptions of the TERESA model were satisfied by the sedimentary conditions in the disturbed Tema Harbour, which provided a unique opportunity to test the performance of this new dating tool under conditions where the assumptions of most of the ^{210}Pb-based models fail. Thus, this study may be of interest to the broad scientific community concerned with the investigation of sedimentation conditions and pollution records in harbours and other dynamic systems where SAR values fall beyond the capabilities of the differencing GIS-bathymetry and conventional ^{210}Pb-based dating methods, and the use of sediment traps and [234]Th and [7]Be profiles show serious constraints.

7.2. Materials and methods

7.2.1. Study area

The Tema Harbour is about 70 km west from the outlet of the Volta River in Ghana and there are several minor riverine systems along the Ghanaian shoreline that deliver important loads of suspended particulate matter to the coastal region (Akrasi, 2011). However, there is no direct riverine inflow in the Tema Harbour (Botwe et al., 2017b). The Tema Harbour layout and its operations have been described previously (Botwe et al., 2017a; Botwe et al., 2017b; Botwe et al., 2017c). Briefly, the Tema Harbour is a semi-enclosed coastal harbour situated in the Gulf of Guinea at Tema (Greater Accra, Ghana) with a water area of 1.7 km^2 (Fig. 7.1). It has been in operation since 1962 and is partitioned into a Main Harbour, a Fishing Harbour (Inner and Outer) and a Canoe Basin. The Main Harbour is concerned with shipping operations and has water depths ranging from 7.5 to 11.4 m (mean 8.5 m), a 240 m wide entrance and a breakwater of 4850 m length (Botwe et al., 2017b). The Fishing Harbour and the Canoe Basin are dedicated to the operations of semi-industrial, industrial and artisanal fishing vessels (Botwe et al., 2017c).

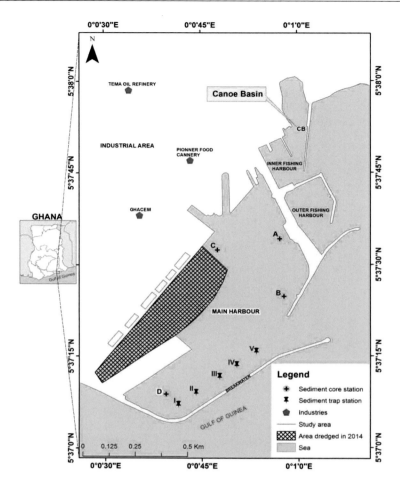

Fig. 7.1 Map of Tema Harbour (Ghana) showing sediment core and sediment trap sampling stations and the approximate area dredged in 2014.

The average water temperature is around 23 °C with water salinity ranging from 30-35‰ (Botwe et al., 2017c). Tides in the study area are semi-diurnal with a tidal range of about 1.6 m. In the immediate coastal sea area, tidal currents range from less than 0.1 to 0.5 m.s^{-1} and wave heights range from 1 to 2 m (http://open_jicareport.jica.go.jp/pdf/11681632_03.pdf). A portion of the Main Harbour (shown in Fig. 7.1) was dredged in March-April 2014, while the Canoe Basin was dredged in May 2013 (Botwe et al., 2017c). It is also worth noting that in 2013, the construction of a new wharf was started near the eastern margin of the dredged area, with the emplacement of the supporting concrete pillars. The over-water structure was completed during 2014.

7.2.2. Sediment sampling

Four sediment cores were sampled with an Uwitech® gravity corer (length = 60 cm; internal diameter = 8.5 cm) from stations A, B, C and D in the Tema Harbour (Fig. 7.1) in April 2015. Water depths at the sediment core stations A, B, C and D were 8.5, 8, 10 and 9 m, respectively. During the sediment core sampling period, the turbidity of seawater was measured in Nephelometric Turbidity Unit (NTU) at the surface, mid depth and the bottom at each core sampling station using a turbidity probe (HI 9829, Hanna Instruments, USA).

Besides, vertical arrays of 2 cylindrical polyvinyl chloride (PVC) sediment traps, positioned 0.6 m apart in an alternating fashion, were deployed at 5 stations (I, II, III, IV and V) along a transect in a less busy area of the Tema Harbour (Fig. 7.1) over a period of 12 weeks (23 May - 15 August, 2015) to collect settling particles at two water depths of approximately 1.8 m (top trap) and 0.6 m (bottom trap) from the seabed. The aspect ratio of each sediment trap was 6.0 (i.e. height = 60 cm; diameter = 10 cm) to minimise current-induced resuspension and loss of material from the trap (Bloesch & Burns, 1980; Kelderman et al., 2012; de Vicente et al., 2010). The sediment traps were retrieved every two weeks (Kelderman, 2012; Kelderman et al., 2012; de Vicente et al., 2010) till the end of the 12-week period. Upon their retrieval, the sediment cores and trap samples were placed upright in racks and transported to the Department of Marine and Fisheries Sciences laboratory at the University of Ghana for further processing and analysis. Rainfall data over the study period was obtained from the Ghana Meteorological Services.

In the laboratory, the sediment cores were allowed to stand overnight, after which the overlying water was carefully siphoned off. Each sediment core was sectioned at 1 cm intervals over the top 2 cm layer and subsequently at 2 cm intervals. During core extrusion and sectioning, lower sections may become contaminated as they move past smears from upper sections left on the walls of the corer. Thus, the outer 1 cm rim of each sediment slice was trimmed off (Bellucci et al., 2012). The content of each sediment trap was wet-sieved successively over 2000 and 63 μm stainless steel mesh sieves to determine the distribution of the <63 μm (silt-clay) and 63-2000 μm (sand) fractions of the trapped sediment. The wet weights of the sediment core samples, the core layers, as well as the <2000 μm fraction of sediment trap samples were measured and then oven-dried at 50°C till constant weight to obtain their dry weights (Botwe et al., 2017b). Direct gamma spectrometric analyses of [7]Be (478 keV), [234]Th (63.3 keV), [210]Pb ([210]Pb$_{tot}$, 46.5 keV), [212]Pb (239 keV), [214]Pb (352 keV), [214]Bi (609 keV), [137]Cs (662 keV) and [40]K (1460 keV) were performed at ENEA Marine Environment Research Centre (S. Teresa, Italy). Prior to the gamma analyses, the sediment core samples were homogenised by grinding and stored air-tight for at least 22 days in plastic vials of standard geometries (5, 10 and 20 g) following Botwe et al. (2017b) to ensure secular equilibrium between the parent nuclides and their short-lived daughter nuclides. The measured activities were decay-corrected to the sediment sampling date.

7.2.3. Data treatment

Sediment trap-derived settling fluxes (SF, with units of g.cm^{-2}.y^{-1}) were estimated from Eq. 1:

$$SF = \frac{M}{A \times D} \qquad (1)$$

where M is the dry accumulated mass of trapped material (g), A is the cross-sectional area of the sediment trap (cm^2) and D is the duration of trap deployment (y).

The water content of the sediment samples was obtained from the difference between the wet and dry weights and expressed as a percentage of the sediment wet weight. The [226]Ra activities were indirectly obtained from the average [214]Pb and [214]Bi activities (Botwe et al., 2017b). For each sediment layer, the [210]Pb$_{exc}$ specific activity was obtained by subtracting the [226]Ra (supported ^{210}Pb) specific activity from the total [210]Pb ([210]Pb$_{tot}$) specific activity (Corbett et al., 2009). Excess [234]Th ([234]Th$_{exc}$) specific activity was estimated from the total [234]Th specific activity by subtracting the background level (the averaged value of concentrations measured downcore, except the topmost layers). Core D required a more complex procedure, since the values measured downcore were not constant. The supported [234]Th fraction near the surface was estimated by assuming a constant [238]U/[226]Ra ratio along the whole core. Bulk densities were evaluated from the determination of the dry weight of the mass contained in a well controlled bulk volume of the sediment, and the mass depths at the centre and at the bottom of each sediment slice determined following Baskaran et al. (2014).

Subsurface sediments may undergo compaction and introduce errors in the determination of sediment depth and age relationships and hence, the estimation of linear sedimentation rates

(Abril, 2003; Lu, 2007). Therefore, the activity profiles of ^{7}Be, ^{234}Th$_{exc}$ and ^{210}Pb$_{exc}$ were related to the mass-depth (mass of sediment per unit area, g.cm^{-2}) instead of the linear-depth of the sediment (Erten, 1997; Lu, 2007). Inventories of ^{7}Be were determined for each core by integrating their downcore activities (Giffin and Corbett, 2003; Mullenbach et al., 2004; Palinkas et al., 2005). The simplest SAR modelling approach, defined by the assumptions of constant flux and constant sedimentation rate (CF-CS) without any post-depositional redistribution (Appleby and Oldfield, 1983; Erten, 1997), was applied to the sediment cores to derive very recent (3-8 months) SARs based on the measured ^{7}Be activity profiles.

The TERESA model was applied for the analysis of ^{210}Pb$_{exc}$ versus mass depth profiles. The model fundamentals and its validation against synthetic cores and real data from varved sediments have been described by Abril (2016). In summary, the model stands on the following set of assumptions: (i) ^{210}Pb$_{exc}$ behaves as a particle-associated tracer and new inputs are ideally deposited over the previously existing material; (ii) there is no post-depositional redistribution; (iii) there is continuity of the sequence (i.e., there is not any missing layer by erosion); and (iv) ^{210}Pb$_{exc}$ fluxes are governed by 'horizontal inputs' and thus, there is a statistical correlation between ^{210}Pb$_{exc}$ fluxes and SAR (Abril and Brunskill, 2014). Therefore, for a sediment core which has been sectioned into N slices of mass thickness $\Delta m_{i\,(i\,=\,1,\,2,\,...\,N)}$, each slice has an associated age interval ΔT_i, a mean SAR value (w_i), and an associated initial specific activity ($A_{0,i}$) corresponding to the activity when that sediment slice was at the sediment-water-interface (SWI).

The TERESA model operates with SAR and initial radionuclide activity distributions, for respectively, w_i and $A_{0,i}$, which closely follow normal distributions around their arithmetic mean values, $\overline{w}$ and $\bar{A}_0$, with standard deviations σ_w and σ_A, respectively. S_w and S_A are, respectively, their normalised values (standard deviation over mean values). Provided a first estimation of $\overline{w}$, $\bar{A}_0$, σ_w and σ_A, the model generates independent random distributions for w_i and $A_{0,i}$; and then an intelligent algorithm solves their best arrangement downcore to fit the experimental ^{210}Pb$_{exc}$ versus the mass depth profile, generating then solutions for the chronological line and for the histories of SAR and sediment fluxes (Abril, 2016). As the result depends on the first estimation of $\overline{w}$, $\bar{A}_0$, σ_w and σ_A, the model applies a mapping technique by iterating the whole process for each parameter varying over a wide range. The error function, Q^2, measures the overall quality of the fit for each individual run of the model (Abril, 2016):

$$Q^2 = \sum_{i=1}^{N} \frac{(A_{th,i} - A_i)^2}{\sigma_i^2} \quad ; \chi^2 = Q^2 / f \qquad (2)$$

where A_i and σ_i are, respectively, the measured value and the analytical error of the radionuclide specific activity at the slice with index i, $A_{th,i}$ being the corresponding value estimated by the model, χ is a measure of the mean distance between the theoretical and experimental profiles in terms of the size of the associated uncertainties, and f is the number of degrees of freedom.

Parametric maps of the χ-function serve to find out the best solution and to support error estimates (Abril, 2016). The typical fundamentals can be applied for error estimates in the four entry parameters (namely $\bar{w}$, σ_w, $\bar{A}_0$, σ_A) through the curvature of the parametric lines in the Q^2 function around the position of the absolute minimum (Bevington and Robinson, 2003). The associated uncertainty in the age at the bottom of each layer, T_j, can then be estimated through the general propagation law:

$$\sigma^2(T_j) = \left(\frac{\partial T_j}{\partial A_m}\sigma(A_m)\right)^2 + \left(\frac{\partial T_j}{\partial w_m}\sigma(w_m)\right)^2 + \left(\frac{\partial T_j}{\partial \sigma_A}\sigma(\sigma_A)\right)^2 + \left(\frac{\partial T_j}{\partial \sigma_w}\sigma(\sigma_w)\right)^2 \qquad (3)$$

Optionally, the model's answers can be better constrained through the use of time markers, when available. As the model-algorithm requires a continuous profile, data from the non-measured sediment slices have been estimated by linear interpolation with ascribed double error-bars to minimise their relative weight in the fit. To avoid the generation of negative values or influencing points when exploring large values of S_w and S_A, lower threshold limits of 0.1 $\bar{A}_0$ and 0.2 $\bar{w}$ have been imposed in all the calculations.

7.3. Results and discussion

7.3.1. Estimates of settling fluxes of particulate matter in the Tema Harbour

Table 7.1 shows the estimated average settling fluxes (SFs) of particulate matter over two sampling depths (top and bottom traps) in the Tema Harbour based on the accumulated dry mass in the sediment traps retrieved every two weeks till the end of the 12-week sampling period. The average SF values over the two sampling depths ranged from 15 ($\pm$ 3) to 54 ($\pm$ 18) g.cm^{-2}.y^{-1}, being higher at stations III and IV. About 93% of these SFs (range 84%-96%) correspond to the fraction of particle sizes larger than 63 μm (Table 7.1). As the typical settling velocities for fine sand and larger particle sizes are over 10 m.h^{-1} (Eisma, 1993; Ji, 2008), they will be quickly removed from the water column.

Table 7.1 Estimated settling fluxes (SFs) of particulate matter (mean ± standard error) at five sampling stations in the Tema Harbour (Ghana) based on biweekly accumulated dry mass in sediment traps deployed at two water depths (top and bottom traps) over a period of 12 weeks. The weight fraction (%) of the trapped sediment samples with Φ<63 μm are also presented. See sampling stations in Fig. 7.1.

Sampling station	Sediment trap deployment	SF (g.cm^{-2}.y^{-1})	Fraction with Φ< 63 μm (%)
I	Top trap[1]	18 ± 11	8.1
	Bottom trap[2]	21 ± 7	7.6
II	Top trap[2]	25 ± 5	9.4
	Bottom trap[2]	15 ± 3	8.7
III	Top trap[2]	46 ± 16	3.8
	Bottom trap[2]	42 ± 12	3.9
IV	Top trap[2]	54 ± 18	4.5
	Bottom trap[2]	23 ± 8	16.1
V	Top trap[2]	30 ± 13	6.3
	Bottom trap[2]	28 ± 9	5.2

[1] $n = 3$; [2] $n = 6$

From the maximum tidal range of 1.6 m (http://www.tides4fishing.com/af/ghana/tema#_tide_table) and the geometry of the Main Harbour entrance (i.e. 240 m width and 8.5 m mean water depth), it is possible to estimate a maximum cross-sectional averaged water current of about 6 cm.s^{-1} at the Main Harbour entrance, and a bottom shear stress slightly over 0.07 Pa by applying standard hydrodynamic principles (Periáñez and Abril, 2014; Abril and Periáñez, 2016). This estimated shear stress, which is associated with maximum tidal currents at the Tema Harbour, is over the threshold value (0.06 Pa) for resuspending clays and fine silts (Eisma, 1993; Ji, 2008). But in the inner area of the Main Harbour, the tidal currents are lower and resuspension of sediments is not expected to occur because of the tidal forcing. Nevertheless, ship traffic and particularly the manoeuvres of big cargos and drilling ships can cause remobilisation of large amounts of sediments, as it can be seen from available aerial photographs (e.g. 10/08/2015; Google Earth). Thus, coexistence of sediment deposition and resuspension is expected to occur. Consequently, the obtained SFs may overestimate the SAR values since the material settled within the sediment traps is not allowed to undergo resuspension, which could occur in real sediments. Thus, the estimated time-averaged SFs from the top and bottom sediment traps at each station represent only a proxy and an upper bound of the expected annually-averaged values at their locations.

Concerning temporal variations, some relative maxima were found at different periods for each station (not shown). The SF values from the top sediment traps were usually comparable or higher than those from the bottom traps, although some exceptions were found at stations I and V. The observed maxima found at different periods for each station did not correspond to a single event of entry of a high mass inflow into the Main Harbour. Moreover, maxima of SFs at the two sampling depths at each sampling station are often registered at different times. This fact points out the importance of local disturbances, which have a highly irregular character both in time and space, and are most likely linked to the manoeuvring of cargo ships. Moreover, from the location of the sampling points, it is expected that station I, at a corner of the Main Harbour, should experience less perturbations and thus capturing lower masses of SF, as indeed found in Table 7.1. This reveals a complex dynamics of rising clouds of suspended particulate matter (SPM) and settling in the Tema Harbour, interfered by horizontal transport, most likely the one forced by the tidal dynamics.

The mean (± standard deviation) turbidity values of the surface, mid depth and bottom seawater at the sediment core sampling stations were as follows: core station A = 3.9 (± 0.8) Nephelometric Turbidity Unit (NTU); B = 3 (± 0.6) NTU; C = 3.7 (± 0.7) and D = 4.7 (± 1.1). Although a station-specific ppm (mg.L^{-1}) versus NTU calibration curve was not constructed, from a typical slope of 3.4 ppm/NTU, the turbidity measurements lead to an estimation of the SPM concentration, C, of around 10 parts per million (ppm). Assuming that this value is representative of the mean environmental conditions at the Tema Harbour and using the SF values in Table 7.1 (i.e. $SF = v_s C$), the mean values for the settling velocities, v_s, were estimated to be in the range of $(0.5\text{-}1.7)10^{-3}$ m.s^{-1}, again corresponding to silt and fine sand fractions. These considerations could also provide a gross estimate of the order of magnitude of the expected mean sedimentation rates in the Tema Harbour. From the known tidal range and mean water depth, one could estimate that in each tidal cycle, the Tema Harbour exchanges about 10% of its water volume. If the above SPM concentration of 10 ppm is representative of the marine inflow and the Tema Harbour acts as a huge sediment trap, the expected mean SAR value would be about 6 g.cm^{-2}.y^{-1}. Large local variability is expected around this value due to localised discharge of solids, ship traffic and varying sedimentation yields.

It is worth noting that the SF values from Table 7.1 captured the prevailing environmental conditions in the SW area of the Tema Harbour during the 12-week deployment period. They represent an upper bound for the sediment mass accumulation rates (SARs) at their respective sampling stations. Rainfall in the sampling area during the sampling period and the preceding month ranged from 0 to 142.3 mm with a mean of 6.1 mm (not shown). There were no episodic incidents of heavy rainfall, storm runoff or floods during the study period, which could have induced sediment transport into the Tema Harbour and affect the sedimentary regime (Palinkas et al., 2005; Díaz-Asencio et al., 2011).

7.3.2. Radionuclide and bulk density profiles in sediment cores from Tema Harbour

The mass depth and ^{7}Be, ^{234}Th, ^{210}Pb, ^{226}Ra, ^{212}Pb and ^{137}Cs specific activity profiles of cores A, B, C and D from the Tema Harbour are shown in Table 7.2, while Fig. 7.2 shows the bulk density versus depth profiles for the cores. Table 7.2 shows that cores A, B and C had similar radionuclide trends, with almost constant downcore ^{210}Pb$_{tot}$ specific activity (around 200-300 Bq.kg^{-1}), while core D showed large differences in the radionuclide activity profiles with lower activities at intermediate depths (from approximately 15 to 25 cm). For all the cores, the ^{210}Pb$_{exc}$ and ^{137}Cs specific activity profiles did not reach a zero value in the deeper layers. Moreover, no clearly decreasing ^{210}Pb$_{exc}$ profiles were observed in all four sediment cores.

Fig. 7.2 shows that bulk density in cores A and B exhibit similar and typical quasi-steady compaction profiles broken by some intermediate relative peaks. Core C showed a trend of increasing bulk density downcore with anomalous low values at the top 0-2 cm layer. Core D

Table 7.2 Profiles of mass depth and radioactivity (^{7}Be, ^{234}Th, ^{210}Pb$_{tot}$, ^{226}Ra, ^{212}Pb, ^{40}K and ^{137}Cs) concentrations in sediment cores from Tema Harbour (Ghana). See sampling stations in Fig. 7.1.

Core ID	Core depth (cm)	*Cumulative mass depth (g.cm^2)	Specific activities (Bq kg^{-1})						
			^{7}Be	^{234}Th	^{210}Pb$_{tot}$	^{226}Ra	^{212}Pb	^{40}K	^{137}Cs
A	0-1	0.65	50 ± 9	135 ± 34	296 ± 13	15 ± 2	40 ± 1	317 ± 18	< 0.6
	1-2	1.27	29 ± 7	94 ± 32	302 ± 10	12 ± 2	41 ± 1	304 ± 17	< 0.6
	2-4	2.39	20 ± 6	28 ± 6	279 ± 10	13 ± 2	36 ± 1	288 ± 15	1.9 ± 0.5
	4-6	4.28	6 ± 3	32 ± 4	319 ± 10	15 ± 1	42 ± 1	313 ± 13	2.1 ± 0.4
	6-8	5.45	5 ± 3	36 ± 4	286 ± 9	14 ± 1	44 ± 1	333 ± 14	1.4 ± 0.4
	8-10	6.62	< 6	35 ± 4	251 ± 7	16 ± 1	47 ± 1	336 ± 14	2.0 ± 0.3
	10-12	7.69	< 6	31 ± 4	314 ± 9	15 ± 1	28 ± 1	294 ± 11	1.0 ± 0.3
	12-14	9.13	< 6	35 ± 3	309 ± 10	15 ± 1	26 ± 1	309 ± 10	1.6 ± 0.3
	16-18	12.67	< 6	30 ± 4	300 ± 9	15 ± 1	25 ± 1	271 ± 12	1.7 ± 0.4
	24-26	17.30	< 6	39 ± 3	330 ± 9	17 ± 1	25 ± 1	292 ± 10	1.4 ± 0.2
	32-34	22.13	< 6	33 ± 3	282 ± 7	15 ± 1	25 ± 1	294 ± 9	1.5 ± 0.3
B	0-1	0.55	61 ± 14	206 ± 30	285 ± 20	11 ± 2	46 ± 3	360 ± 31	<1
	1-2	1.34	17 ± 9	84 ± 46	233 ± 17	13 ± 2	45 ± 2	329 ± 27	2.6 ± 1.0
	2-4	2.71	<9	32 ± 6	245 ± 14	14 ± 1	43 ± 1	323 ± 17	2.1 ± 0.6
	4-6	3.91	<9	43 ± 6	298 ± 15	14 ± 1	29 ± 1	291 ± 15	1.8 ± 0.5
	6-8	5.1	<9	34 ± 9	298 ± 18	16 ± 2	29 ± 1	295 ± 21	1.8 ± 0.7
	8-10	7.3	<9	35 ± 6	284 ± 14	14 ± 1	26 ± 1	285 ± 12	2.0 ± 0.4
	12-14	10.0	<9	40 ± 6	298 ± 15	14 ± 1	27 ± 1	326 ± 15	1.7 ± 0.4
	16-18	12.2	<9	40 ± 6	222 ± 12	13 ± 1	28 ± 1	320 ± 17	0.9 ± 0.4
	22-24	16.6	<9	34 ± 5	175 ± 10	14 ± 1	37 ± 1	324 ± 13	1.8 ± 0.3
	28-30	20.8	<9	29 ± 5	165 ± 10	15 ± 1	37 ± 1	340 ± 14	2.2 ± 0.5
C	0-1	0.26	55 ± 10	196 ± 19	258 ± 17	10 ± 2	48 ± 2	325 ± 28	2.1 ± 0.9
	1-2	0.65	54 ± 12	278 ± 25	276 ± 17	13 ± 2	49 ± 2	375 ± 28	<1
	2-4	2.06	25 ± 5	95 ± 16	242 ± 14	13 ± 1	42 ± 1	314 ± 11	1.2 ± 0.4
	4-6	3.39	<12	36 ± 3	289 ± 16	14 ± 1	43 ± 1	341 ± 11	1.1 ± 0.3
	6-8	4.74	<12	32 ± 4	302 ± 18	14 ± 1	45 ± 1	341 ± 12	1.0 ± 0.4
	8-10	6.31	<12	30 ± 3	182 ± 12	15 ± 1	60 ± 1	396 ± 13	1.3 ± 0.4
	12-14	9.92	<12	34 ± 5	177 ± 13	15 ± 1	59 ± 1	373 ± 15	1.1 ± 0.5
	14-16	11.89	<12	31 ± 5	187 ± 14	15 ± 1	53 ± 1	406 ± 17	1.9 ± 0.5
	16-18	14.14	<12	28 ± 4	155 ± 10	15 ± 1	51 ± 1	453 ± 13	<1
	18-20	16.10	<12	37 ± 4	190 ± 12	17 ± 1	62 ± 1	413 ± 15	1.0 ± 0.3
	32-34	30.95	<12	37 ± 3	203 ± 12	13 ± 1	34 ± 1	404 ± 12	1.3 ± 0.3
	40-42	41.10	<12	37 ± 4	219 ± 13	14 ± 1	35 ± 1	402 ± 12	1.7 ± 0.5
	46-48	48.8	<12	31 ± 5	207 ± 10	16 ± 1	32 ± 1	421 ± 15	1.5 ± 0.3

Table 7.2 Continued

Core ID	Core depth (cm)	*Cumulative mass depth (g.cm^2)	Specific activities (Bq.kg^{-1})						
			^{7}Be	^{234}Th	^{210}Pb$_{tot}$	^{226}Ra	^{212}Pb	^{40}K	^{137}Cs
D	0-1	0.38	126 ± 31	219 ± 37	276 ± 16	17 ± 3	39 ± 1	333 ± 19	2.4 ± 0.8
	1-2	0.94	48 ± 7	276 ± 38	284 ± 13	15 ± 2	40 ± 1	340 ± 19	<1.2
	2-4	2.12	34 ± 9	234 ± 40	269 ± 13	11 ± 2	37 ± 1	321 ± 18	<1.2
	4-6	3.27	19 ± 5	107 ± 15	242 ± 10	12 ± 2	35 ± 1	299 ± 14	<1.2
	6-8	4.82	<6	47 ± 14	217 ± 13	9 ± 2	28 ± 1	302 ± 18	<1.2
	8-10	6.41	<6	25 ± 3	200 ± 8	11 ± 2	25 ± 1	270 ± 13	1.3 ± 0.4
	12-14	10.87	<6	16 ± 3	82 ± 6	7 ± 1	16 ± 0.5	266 ± 11	<0.6
	16-18	17.22	<6	7 ± 2	18 ± 4	5 ± 1	11 ± 0.3	275 ± 9	<0.6
	20-22	24.33	<6	9 ± 3	41 ± 6	4 ± 1	14 ± 0.4	318 ± 13	0.6 ± 0.3
	26-28	33.85	<6	17 ± 2	79 ± 5	9 ± 1	21 ± 0.4	280 ± 10	0.8 ± 0.2
	36-38	48.10	<6	13 ± 2	86 ± 6	7 ± 1	17 ± 0.4	299 ± 11	<0.6
	38-40	49.73	<6	24 ± 4	182 ± 7	12 ± 1	28 ± 0.6	337 ± 12	1.0 ± 0.3

Errors are 1 standard deviation from counting statistics; Minimum Detectable Activity are indicated as < MDA (Bq.kg^{-1}); *At the central point of each sediment slice

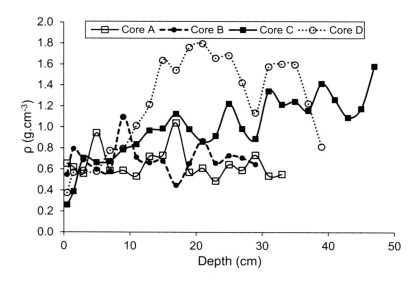

Fig. 7.2 Bulk density (ρ) versus depth profiles for sediment cores A, B, C and D sampled from Tema Harbour (Ghana)

showed a distinct structure, with higher values of bulk density below the 15 cm depth. Thus, the data on bulk density suggest highly dynamic and non-steady sedimentary conditions at the study site (Abril, 2003). It is worth noting that high values of bulk density are typically associated with the presence of coarser grain-sizes, which have lower specific activities of surface-bound radionuclides.

7.3.3. Inventories and fluxes of ^{7}Be and ^{7}Be-derived sediment accumulation rates

The measurement of ^{7}Be specific activities in the top layer of a sediment core is usually considered as a quality test for the complete surface recovery of the core, which is necessary for dating (Erten et al., 1985; Erten, 1997). In the Tema Harbour, it was possible to measure ^{7}Be in several sediment slices downcore (Table 7.2). This provided an opportunity to determine the very recent SAR values over the past 6-8 months under some simplifying assumptions. The ^{7}Be inventories (see Table 7.3) ranged from 470 (± 110) to 1360 (± 170) Bq.m^{-2}, core D being richer in ^{7}Be than the other cores. The ^{7}Be inventories in the Tema Harbour sediments were higher, but comparable to the values reported by Olsen et al. (1985) for coastal sediments from Virginia and Tennessee (370-740 Bq.m^{-2}). The ^{7}Be inventories in the Tema Harbour sediments were also comparable to, but lower than the values reported for

(1) sediments of the Neuse and Pamlico River Estuaries (33-97 $Bq.m^{-2}$) by Giffin and Corbett (2003), (2) sediments of the Eel Canyon, northern California (5,780-12,000 $Bq.m^{-2}$) by Mullenbach et al. (2004), and (3) sediments from Po River delta, Adriatic Sea (190-5,420 $Bq.m^{-2}$) by Palinkas et al. (2005). The fluxes of ^{7}Be onto the SWI, assumed to be constant, were estimated from the product of the ^{7}Be inventories and the radioactive decay constant. They varied from 2200 ($\pm$ 500) to 6500 ($\pm$ 800) $Bq.m^{-2}.y^{-1}$ (Table 7.3). The spatial variability in ^{7}Be values in the Tema Harbour may be due to the fact that uptake of atmospheric ^{7}Be by SPM mediates its fluxes onto the SWI. A large spatial variability in ^{7}Be inventories is a common feature of coastal sediments, which has been previously reported in the scientific literature (Rose and Kuehl, 2010; Kolker et al., 2012).

Table 7.3 ^{7}Be inventories and fluxes onto the sediment-water interface (SWI) and ^{7}Be-derived sediment accumulation rate (SAR) values in sediment cores From Tema Harbour (Ghana).

Core	^{7}Be inventory (Bq.m^{-2})	^{7}Be flux onto SWI (Bq.m^{-2}.y^{-1})	^{7}Be-derived SAR* (g.cm^{-2}.y^{-1})
A	900 ± 120	4300 ± 600	9.0 ± 1.3
B	470 ± 110	2200 ± 500	2.5 ± nd
C	710 ± 90	3400 ± 400	7.1 ± 1.8
D	1360 ± 170	6500 ± 800	7.1 ± 2.0

*From a CF-CS model; nd: not determined

Fig. 7.3 shows that ^{7}Be versus mass depth profiles in sediment cores A and D followed an exponential decay at a confidence level of over 90%, while for cores B and C the number of data points is not statistically significant. Table 7.3 reports the ^{7}Be-derived SAR values, assuming the applicability of the CF-CS model. The SAR values were in the range of 2.5-9.0 $g.cm^{-2}.y^{-1}$, and were much lower than the sediment trap-derived SFs (Table 7.1). This observation points out the important role of sediment resuspension in the sedimentation process at the Tema Harbour, as expected from the aerial photographs (see Section 7.3.1). Nevertheless, these values have to be taken as a first estimate of the order of magnitude of SARs in Tema Harbour. The reported uncertainties in Table 7.3 come from a numerical fitting which does not account for the uncertainties in the radionuclide specific activities. But "the model errors" arising from a partial accomplishment of the involved assumptions may be more relevant. Thus, the ^{7}Be versus mass-depth profiles could be mediated by mixing, diffusion or non-ideal depth deposition of fluxes, which would result in somewhat lower SAR values than the above estimates. Particularly, the assumption of a constant flux is problematic for [7]Be due to its short half-life (Taylor et al., 2013).

It has been shown for different geographical latitudes that the atmospheric deposition of [7]Be is well correlated with rainfall (Pham et al., 2013). Rainfall in southern Ghana is higher during the rainy season (April-June and September-October) and minimum during the dry season (November-March and July-August). Unusually high rainfall values of about 100 mm were recorded for both February and March 2015, while only 30-40 mm was recorded in January 2015 (https://jbaidoowilliams.com/2015/10/20/el-nino-2015-rainfall-fears/). Although a

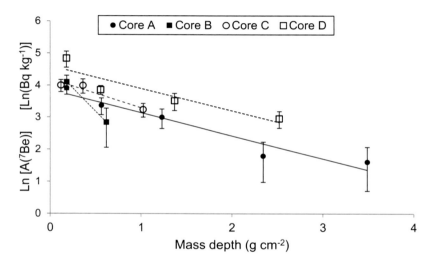

Fig. 7.3 [7]Be versus mass depth profiles of sediment cores A, B, C and D sampled from Tema Harbour (Ghana). Error bars represent 1-σ counting uncertainties.

system time-averaged integration of fluxes has been described (Robbins, 2000), the assumption of constant fluxes of [7]Be may be an oversimplification. On the other hand, the short penetration depth of [7]Be in the core makes its profile more sensitive to surficial processes such as the non-ideal deposition described by Abril and Gharbi (2012).

The [234]Th activity versus mass depth profiles (Table 7.2) showed a near-constant background level along the cores at the 2-3 cm upper sediment slices. This is consistent with the high SAR values estimated from the [7]Be data.

7.3.4. Analysis of $^{210}Pb_{exc}$ fluxes onto the SWI and mixing

From Table 7.2, the surficial specific activities of $^{210}Pb_{exc}$ (~300 Bq.kg^{-1}) were over one order of magnitude higher than those reported by Nyarko et al. (2016) for sediment cores from the Pra and Volta estuaries in Ghana. The difference can be explained in terms of the granulometry of the sediments, with a large component of the inputs linked to the small grain-size SPM supplied by the tidal inflow. For the theoretical basis of the relationship among radionuclide specific activities and particle size, see Abril and Fraga (1996) and Abril (1998). The analysed cores were too short to allow any reliable estimation of the total $^{210}Pb_{exc}$ inventories required for the application of the constant rate of supply (CRS) model. The low values and large uncertainties in the ^{137}Cs data did not allow any proper identification of chronostratigraphic horizons. Moreover, cores A, C and D did not follow any clear monotonic exponential trend of decrease (Fig. 7.4) and thus, the application of the CF-CS model for deriving SAR values was not reliable.

The SAR values derived from ^{7}Be data, along with the $^{210}Pb_{exc}$ specific activities in the upper sediment layers can be used to obtain a first estimate of the $^{210}Pb_{exc}$ fluxes onto the SWI if any post-depositional remobilization is discarded. Thus, for core A, this resulted in a flux onto the SWI of 25.6 kBq.m^{-2}.y^{-1}, being two orders of magnitude higher than the expected atmospheric deposition of $^{210}Pb_{exc}$ in the Tema Harbour area (typically in the range of 100-200 Bq.m^{-2}.y^{-1}). This means that fluxes of $^{210}Pb_{exc}$ onto the SWI must be governed by fluxes of matter (Abril and Brunskill, 2014). Furthermore, the models handling high diffusion coefficients and null or low SAR values under steady-state conditions could roughly fit the profiles from cores A, B and C (not shown). However, these models require fluxes of up to 100 kBq.m^{-2}.y^{-1} to enter into the SWI (mostly in the dissolved form), which cannot be supplied by atmospheric deposition.

About mixing, on the other hand, no observational evidence of bioturbation was available that would induce sediment mixing over large mass depths. Some degree of horizontal and vertical mixing may, nevertheless, exist due to the remobilization forced by ship displacements. But as data from sediment traps integrate a large number of such events, mixing in the sediments is not expected to exceed the 1-2 cm surface layer. About 70 vessels enter and leave the Tema Harbour weekly, most of them being cargo, tug and tanker vessels; and they follow well-defined   tracks   within   the   harbour (http://www.marinetraffic.com/en/ais/details/ports/86/Ghana_port:TEMA). In contrast, the depth profiles of ^{7}Be and ^{234}Th in the sampling sites showed sharp gradients in the 0-2 cm layers, which are not compatible with a well-mixed layer. The hypothesis of mixing can thus be discarded as a main governing factor to explain the observed profiles. As the cores were sampled out of the dredged area (Fig. 7.1), they must have preserved a sequence of continuous sediment deposition.

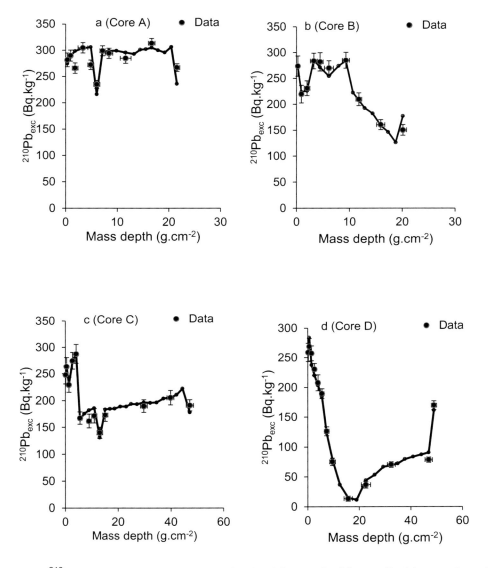

Fig. 7.4 $^{210}Pb_{exc}$ versus mass depth profile for (a) core A, (b) core B, (c) core C and (d) core D in the Tema Harbour (Ghana). Vertical bars correspond to the associated uncertainties (1-σ), while the horizontal ones define the mass depth interval of each sediment slice. The numerical solution generated by the TERESA model is plotted as points at the centre of each slice interval (continuous line is only for guiding-eyes).

7.3.5. Analyses of ^{210}Pb profiles and estimation of SARs using the TERESA model

The hypothesis of varying, but statistically correlated, fluxes of matter and ^{210}Pb$_{exc}$ onto the SWI seems reasonable for the Tema Harbour and thus, the conditions for applying the TERESA model can be met. Maps for the χ function (described in Section 7.2.3) in the ($\overline{A}_0$, $\overline{w}$) and (S_A - S_W) spaces for the four studied cores were computed with the TERESA model (not shown). The computation was done with the stand-alone version of the model and with the basic method-A (Abril, 2016), and they show relatively well defined valleys which determine the parameter values along with their associated fitting uncertainties, reported in Table 7.4. The computed ^{210}Pb$_{exc}$ vs. mass depth profiles are shown in Fig. 7.4(a-d), and they match the experimental data with the χ values reported in Table 7.4. Finally, the resulting chronologies for the four cores are shown in Fig. 7.5.

Table 7.4 Entry parameters of the TERESA model and fitting-error estimates[1] for the investigated cores from the Tema Harbour, where $\overline{A}_0$ (Bq.kg^{-1}) is the mean initial concentration, $\overline{w}$ is the mean SAR value (g.cm^{-2}.y^{-1}) and S_A, S_w are the corresponding relative standard deviations.

Core	$\overline{A}_0$	$\overline{w}$	S_A	S_w	χ
A	364.0 ± 0.2	1.620 ± 0.008	0.175 ± 0.003	0.384 ± 0.006	1.90
B	262.0 ± 1.0	1.87 ± 0.07	0.169 ± 0.009	0.322 ± 0.024	0.96
C	253.0 ± 1.2	3.27 ± 0.21	0.200 ± 0.010	0.27 ± 0.08	0.61
D	152.5 ± 0.1	4.080 ± 0.004	0.600 ± 0.001	0.990 ± 0.003	1.09

[1] Through the second derivative of the Q^2 function (Bevington and Robinson, 2003)

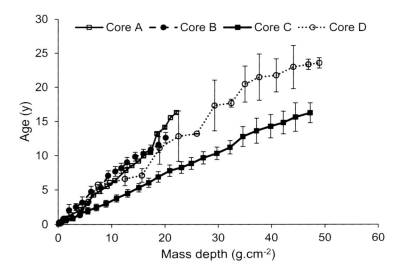

Fig. 7.5 Chronologies estimated by the TERESA model for the investigated sediment cores from the Tema Harbour (Ghana) along with their corresponding uncertainty intervals (from the propagation of error associated to the fitting parameters of Table 7.4).

7.3.6. Analyses of sediment cores A, B and C

Data from sediment core A showed the highest values of ^{210}Pbexc specific activities and they remained almost constant downcore. The TERESA model produced a reasonable fit to the data with arithmetic means of 364 Bq kg^{-1} and 1.62 g.cm^{-2}.y^{-1} and normalised standard deviations of 0.175 and 0.384 for the initial specific activity and SAR, respectively. It is worth noting that there were no alternative valley regions in the χ surface, and the fit, although not especially good, was well constrained, as inferred from the low fitting uncertainties (which are related to the curvature of the χ-surface along the parametric lines). There was a trend of decreasing radionuclide specific activities in recent dates, which compensates for the radioactive decay in the deeper layers and explains the observed profile. The age of the deepest measured sediment slice (32-34 cm) was estimated at 16 y, which implies a time-averaged SAR of 1.4 g.cm^{-2}.y^{-1} (or 2.1 cm.y^{-1}). The time-averaged SAR is used likewise the CRS dating model in order to make these values comparable with literature data.

The chronological line generated by the TERESA model for sediment core B ran close to the one from core A (Fig. 7.5), with a slightly higher value of SAR and lower initial radionuclide specific activity (arithmetic means of 1.87 g.cm^{-2}.y^{-1} and 262 Bq kg^{-1}, respectively). The age of the deepest measured sediment slice (28-30 cm) was estimated at 13 y, which implies a time-averaged SAR of 1.6 g.cm^{-2}.y^{-1} (or 2.4 cm.y^{-1}).

Sediment core C showed a ^{210}Pb$_{exc}$ profile with a distinct sub-surface peak followed by a plateau area, and it was well reproduced by the TERESA model (Fig. 7.4c). The arithmetic mean value for the initial specific activity was close to the one for sediment core B, but the SAR value was noticeably higher. The age of the deepest measured sediment slice (46-48 cm) was estimated at 16 y (with a time-averaged SAR of 2.9 g.cm^{-2}.y^{-1} or 3.0 cm.y^{-1}). The distinct peak extended up to a mass-depth of 4.06 g.cm^{-2}, for which the TERESA-derived age was 1.5 ($\pm$ 0.5) y (Fig. 7.5). The building of the wharf structure between 2013 and 2014 might have changed the local sedimentological conditions. Under the new conditions, the radionuclides content in sediment core C tends to the values found in sediment core A (see Table 7.2). This implies that before the deployment of the pillars, the fluxes of matter on the site of sediment core C were contributed by both marine inflow and some local sources, which resulted in higher bulk densities (Fig. 7.2), higher SAR (Table 7.4), lower ^{210}Pb$_{exc}$ activity content and some higher ^{40}K specific activities (see Table 7.2). The sharp discontinuity in radionuclides profiles at the mass-depth of 4.06 g.cm^{-2} confirms the hypothesis of negligible post depositional processes.

7.3.7. Analysis of sediment core D

Sediment core D showed a complex ^{210}Pb$_{exc}$ profile with high specific activities around 270 Bq.kg^{-1} at the top layers, which decline till 13 Bq.kg^{-1} at a mass depth of 17 g.cm^{-2} and then increase downcore till 170 Bq.kg^{-1} (Fig. 7.4d). The area with low ^{210}Pb$_{exc}$ specific activities was associated with high values of bulk density (Fig. 7.2). The TERESA model was able to reasonably reproduce the complex ^{210}Pb$_{exc}$ profile of sediment core D ($\chi = 1.09$, see Fig. 7.4d) with arithmetic means of initial activities and SAR which are roughly 2/5 and 5/2 of the ones found for sediment core A (Table 7.4). A large variability was ascribed to both values with reference normalised standard deviations of 0.60 and 0.99, respectively. The computed initial specific activities was plotted versus mass depths to display a U-shape with a minimum mass depth around 17-21 g.cm^{-2} (age 11-13 y). This induced a huge dilution effect, most likely attributable to local sources of matter with negligible ^{210}Pb$_{exc}$ content, so the mean value of the flux onto the SWI was 6.2 kBq.m^{-2}.y^{-1}, as in sediment core A. As in the case of sediment core C, these local sources also resulted in higher bulk densities (Fig. 7.2) and the overall higher SAR (Table 7.4). The variability in SAR was very high, but randomly distributed, which resulted in a relatively uniform chronological line (Fig. 7.5). The age of the deepest measured sediment slice (38-40 cm) was estimated at 24 y, which implies a time-averaged SAR of 2.1 g.cm^{-2}.y^{-1} (or 1.7 cm.y^{-1}).

The above interpretation can be further supported by the analysis of the ^{234}Th$_{exc}$ profile in sediment core D. When excluding the topmost layers, the ^{234}Th concentrations were almost constant downcore with almost uniform ^{234}Th/^{226}Ra isotopic ratios, having mean ($\pm$ standard deviation) values of 2.24 ($\pm$ 0.17), 2.57 ($\pm$ 0.45), 2.28 ($\pm$ 0.33) and 1.72 ($\pm$ 0.15) for sediment cores A, B, C and D, respectively. When comparing the concentrations of ^{234}Th and ^{226}Ra along the four sediment cores, it is striking to note that in core D, their values were about a factor of 2 lower in the central layers associated with low values of ^{210}Pb$_{exc}$ and high bulk density values. The low radionuclide specific activities, while keeping the isotopic ratios, can only be explained by the dilution of naturally settling material with other sources of matter. A similar effect of dilution is observed for ^{226}Ra and ^{212}Pb isotopes, which show an increasing rend in the top sediment layers (Fig. 7.6). This means that the dilution effect progressively vanished during the last six years, according to the chronology reported in Fig. 7.5. Their concentrations at the recent SWI converge to the ones found for sediment core A (see Table 7.2). The dilution was not so apparent for ^{40}K (Fig. 7.6) and thus this radionuclide should be present in the local source of matter at concentrations somewhat lower than those found in sediment core A.

The reported uncertainties in the entry parameters of the TERESA model (Table 7.4) came from the fitting procedure, while the ones reported with the chronologies (Fig. 7.5) came from their propagation. As with all the models, there are other sources of uncertainty associated with the partial accomplishment of the model assumptions. In the TERESA model, the reference normal distributions for initial activities and SAR (truncated with the minimum threshold values) are only a proxy to the real ones occurring in nature. The incomplete sequence of measurements make interpolations necessary, and the absence of independent and well distributed chronostratigraphic marks did not allow the use of the complete model capabilities (Abril, 2016).

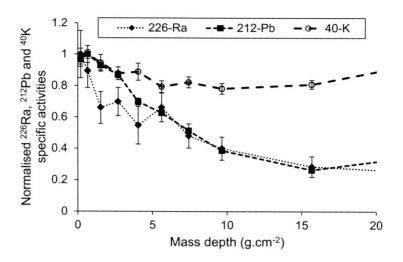

Fig. 7.6 Normalised (to their maximum value) ^{226}Ra, ^{212}Pb and ^{40}K specific activities versus mass depth for sediment core D sampled from the Tema Harbour (Ghana).

7.3.8. Comparison of SAR values among sediment cores A, B, C and D

Despite the above limitations, the TERESA model was able to depict a good scenario for the sedimentation in the Tema Harbour, with overall SAR values with time-averages in the range of 1.4-3.0 g.cm^{-2}.y^{-1}. The analysed sediment core lengths were too short and only captured the last 13-24 years of sedimentation (extreme values for sediment cores B and D, respectively). Sediment cores A and B seemed to capture the less perturbed conditions, and they were governed by the marine inflow of SPM with higher ^{210}Pb$_{exc}$ specific activities and most likely linked to a finer granulometry. The TERESA model captured a change in the sedimentary conditions for sediment core C, whose age is in reasonable agreement with the time of deployment of the concrete pillars of the new wharf structure. According to the model results, the previous sedimentary environment around sediment core C also received matter with lower radionuclide activity content from local sources, leading to high SAR and high bulk densities. A more extreme change in sedimentary conditions was observed for the sediment core D site, which is located in the harbour area transited by cargo ships. Here, the TERESA model reproduced a noticeable dilution effect also attributable to the mixing of the marine inflow of SPM with local sources of matter that had low specific activities of $^{210}Pb_{exc}$, ^{226}Ra, ^{234}Th and ^{212}Pb.

7.3.9. Comparison of the TERESA- and ^{7}Be-derived SARs

The SAR values derived from the TERESA model were of the same order, but noticeably lower than those estimated from the ^{7}Be data. As discussed above, the assumption of a constant flux may be an oversimplification for ^{7}Be and so the SAR was not constant. The overestimation of SAR calculated by ^{7}Be is most likely the major effect of a mass-depth penetration of a fraction of the ^{7}Be fluxes (non-ideal deposition as described by Abril and Gharbi, 2012). A fraction of the radiotracer flux crossing the SWI exists in dissolved form or attached to small grain-size fractions. This fraction can penetrate through the connected water pores and be distributed at depth instead of being deposited over the previously existing material. This effect has been widely observed for ^{7}Be in soils (Taylor et al., 2013), riverine floodplain sediments (Sommerfield et al., 1999), and in marine environments (Rose and Kuehl, 2010). Thus, as a demonstration of the concept, Fig. 7.7 plots the results from a numerical simulation for the ^{7}Be distribution in core A using the SAR value from the TERESA model, and a non-ideal deposition described by factors $g = 0.5$ and $\alpha = 0.6\ g^{-1}.cm^2$ (see Abril and Gharbi, 2012). The plot was compared against the corresponding exponential decline from a CF-CS model with ideal deposition under the same sedimentological conditions. Figure 7.7 confirms non-ideal deposition may seriously limit the applicability of short-lived radionuclides to radiometric dating of recent sediments, while its effects less affected $^{210}Pb_{exc}$: it produced slight distortions in the upper layers, where subsurface maxima are often found (Abril and Gharbi, 2012). The TERESA model compensates for these effects on the $^{210}Pb_{exc}$ profile through an equivalent variability in initial radionuclide activities and/or SAR, as shown in Abril (2016) with the TERESA chronology for a varved sediment core from the Santa Barbara Basin (California, USA) for which non-ideal sediment deposition has been described by Abril and Gharbi (2012).

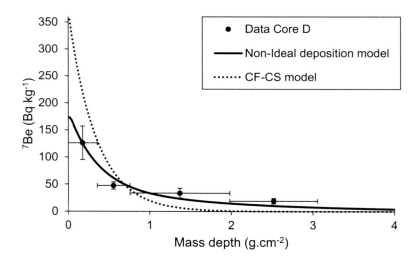

Fig. 7.7 Numerical simulation for the ^{7}Be distribution in sediment core D using the SAR value from the TERESA model, and a non-ideal deposition described by factors $g = 0.5$, $\alpha = 0.6 \ g^{-1}cm^2$ (see Abril and Gharbi, 2012), and the simplifying assumption of constant flux. The plot is compared against the corresponding exponential decline from a CF-CS with ideal deposition under the same sedimentological conditions. Horizontal bars correspond to the depth intervals.

The time-averaged SAR values found in the tropical coastal Tema Harbour by the TERESA model (1.4-3.0 g.cm^{-2}.y^{-1}) were very high when compared with most of the literature data for lacustrine and coastal environments from other climatic regions. Accounting for the high bulk density of the sediments in the Tema Harbour, the accretion rates were in the range of 1.7-3 cm.y^{-1}. Also, these accretion rates were higher than values reported in the scientific literature for harbour areas, e.g. 0.57 cm.y^{-1} in the Tanjung Pelepas Harbour, Malaysia (Yussof et al., 2015); 1.2 cm.y^{-1} in the Victoria Harbour, Hong Kong (Tang et al., 2008); 0.2-2 cm.y^{-1} for Sydney Harbour, Nova Scotia, Canada (Smith et al., 2009). This implies that typical coring devices, allowing for core lengths below 1 m may be too short for recovering the whole ^{210}Pb$_{exc}$ inventory. This, along with irregular non-monotonic profiles, seriously limited the application of most of the conventional ^{210}Pb-based dating models in the studied Tema Harbour cores. The accretion rates in the Tema Harbour are, nevertheless, only of moderate concern for the harbour sustainability, i.e. filling up the harbour basin and imposing a need for dredging. If these conditions had prevailed since its construction in 1962, the net accretion around the studied sites would have been in the range of 0.9-1.6 m, which is in agreement with the partial dredging conducted in 2014.

7.4. Conclusions

Settling fluxes and sedimentation rates in coastal Tema Harbour (Ghana) have been investigated based on accumulated dry mass in sediment traps and radionuclide (^{7}Be, ^{234}Th, ^{210}Pb, ^{226}Ra, ^{40}K, ^{212}Pb and ^{137}Cs) profiles in four sediment cores. The average settling fluxes ranged from 15 ($\pm$ 3) to 54 ($\pm$ 18) g.cm^{-2}.y^{-1}, showing spatial and temporal variations from which a complex dynamics of suspended particulate matter in the Tema Harbour is inferred. The analysed sediment cores were too short to allow any reliable estimation of SARs in the Tema Harbour based on the ^{210}Pb$_{exc}$ inventories and traditional dating models (CF-CS and CRS). The ^{210}Pb-based TERESA model was successfully applied to depict a good scenario for sedimentation rates in the Tema Harbour over the past 13-24 y, with overall time-averaged SAR values in the range of 1.4-3.0 g.cm^{-2}.y^{-1}. These SAR values imply sediment accretion rates of 1.7-3 cm.y^{-1} in the Tema Harbour, which poses a moderate problem of sustainability for its management.

References

Abril JM, Brunskill GJ. Evidence that excess ^{210}Pb flux varies with sediment accumulation rate and implications for dating recent sediments. J Paleolimnol 2014; 52(3):121-137.

Abril JM, Gharbi F. Radiometric dating of recent sediments: beyond the boundary conditions. J Paleolimnol 2012; 48(2):449-460.

Abril JM. A new theoretical treatment of compaction and the advective-diffusive processes in sediments: a reviewed basis for radiometric dating models. J Paleolimnol 2003; 30(4):363-370.

Abril JM. Constraints on the use of ^{137}Cs as a time-marker to support CRS and SIT chronologies. Environ Pollut 2004; 129(1):31-37.

Abril JM, Fraga E. Some physical and chemical features of the variability of K_d distribution coefficients for radionuclides. J Environ Radioact 1996; 30(3):253-270.

Abril JM. Basic microscopic theory of the distribution, transfer and uptake kinetics of dissolved radionuclides by suspended particulate matter - Part I: Theory development. J Environ Radioact 1998; 41(3):307-324.

Abril JM. A ^{210}Pb-based chronological model for recent sediments with random entries of mass and activities: Model development. J Environ Radioact 2016; 151:64-74.

Abril JM, Periáñez R. Revisiting the time scale and size of the Zanclean flood of the Mediterranean (5.33 Ma) from CFD simulations. Mar Geol 2016; 382:242-256.

Akrasi S. Sediment discharges from Ghanaian rivers into the sea. West Afr J Appl Ecol 2011; 18(1):1-13.

Alonso-Hernandez C, Diaz-Asencio M, Muñoz-Caravaca A, Delfanti R, Papucci C, Ferretti O, Crovato C. Recent changes in sedimentation regime in Cienfuegos Bay, Cuba, as inferred from ^{210}Pb and ^{137}Cs vertical profiles. Cont Shelf Res 2006; 26(2):153-167.

Appleby PG, Oldfield F. The assessment of ^{210}Pb data from sites with varying sediment accumulation rates. Hydrobiologia 1983; 103(1):29-35.

Appleby PG, Oldfield F. The calculation of lead-210 dates assuming a constant rate of supply of unsupported ^{210}Pb to the sediment. Catena 1978; 5:1-8.

Barneveld H, Hugtenburg J. Feasibility study for implementation of sedimentation reduction measures in river harbours. RCEM 2008; 1187-1192.

Baskaran M, Nix J, Kuyper C, Karunakara N. Problems with the dating of sediment core using excess ^{210}Pb in a freshwater system impacted by large scale watershed changes. J Environ Radioact 2014; 138:355-363.

Bevington, P. A., Robinson, D.K., 2003. Data Reduction and Error Analysis for the Physical Sciences, 3rd Edition. McGraw-Hill, New York.

Bellucci LG, Giuliani S, Romano S, Albertazzi S, Mugnai C, Frignani M. An integrated approach to the assessment of pollutant delivery chronologies to impacted areas: Hg in the Augusta Bay (Italy). Environ Sci Technol 2012; 46(4):2040-2046.

Bevington PA, Robinson DK. Data Reduction and Error Analysis for the Physical Sciences, 3rd Edition. McGraw-Hill 2003.

Bloesch J, Burns N. A critical review of sedimentation trap technique. Schweiz Z Hydrol 1980; 42(1):15-55.

Botwe BO, Kelderman P, Nyarko E, Lens PNL. Assessment of DDT, HCH and PAH contamination and associated ecotoxicological risks in surface sediments of coastal Tema Harbour (Ghana). Mar Pollut Bull 2017a; 115(1–2):480-488.

Botwe BO, Schirone A, Delbono I, Barsanti M, Delfanti R, Kelderman P, Nyarko E, Lens PNL. Radioactivity concentrations and their radiological significance in sediments of the Tema Harbour (Greater Accra, Ghana). J Radiat Res Appl Sci 2017b; 10(1):63-71.

Botwe BO, De Schamphelaere K, Schipper CA, Teuchies J, Blust R, Nyarko E, Lens PNL. Integrated hazard, risk and impact assessment of tropical marine sediments from Tema Harbour (Ghana). Chemosphere 2017c; 177:24-34.

Brisson LN, Wolfe DA, Staley M. Interferometric swath bathymetry for large scale shallow water hydrographic surveys. Canadian Hydrographic Conference 2014; 1-18.

Brucker S, Clarke JH, Beaudoin J, Lessels C, Czotte K, Loschiavo R, Iwanowska K, Hill P. Monitoring flood-related change in bathymetry and sediment distribution over the Squamish Delta, Howe Sound, British Columbia. U.S. Hydrographic Conference 2007; 1-16.

Buesseler KO, Antia AN, Chen M, Fowler SW, Gardner WD, Gustafsson O, Harada K, Michaels AF, Rutgers van der Loeff M, Sarin M. An assessment of the use of sediment traps for estimating upper ocean particle fluxes. J Mar Res 2007; 65(3):345-416.Caroll J, Lerche I. Sedimentary Processes: Quantification Using Radionuclides. Elsevier 2003.

Corbett DR, Walsh JP, Marciniak K. Temporal and spatial variability of trace metals in sediments of two adjacent tributaries of the Neuse River Estuary, North Carolina, USA. Mar Pollut Bull 2009; 58(11):1739-1747.

de Vicente I, Cruz-Pizarro L, Rueda FJ. Sediment resuspension in two adjacent shallow coastal lakes: controlling factors and consequences on phosphate dynamics. Aquat Sci 2010; 72(1):21-31.

Díaz-Asencio M, Alvarado JC, Alonso-Hernández C, Quejido-Cabezas A, Ruiz-Fernández A, Sanchez-Sanchez M, Hernández-Albernas JI, Eriksson M, Sanchez-Cabeza J. Reconstruction of metal pollution and recent sedimentation processes in Havana Bay (Cuba): a tool for coastal ecosystem management. J Hazard Mater 2011;196:402-411.

Eisma D. Suspended Matter in the Aquatic Environment. Springer-Verlag 1993.

Erten H, Von Gunten H, Rössler E, Sturm M. Dating of sediments from Lake Zurich (Switzerland) with210Pb and137Cs. Swiss J Hydrol 1985; 47(1):5-11.

Erten H. Radiochronology of lake sediments. Pure Appl Chem 1997; 69(1):71-76.

Fuller C, van Geen A, Baskaran M, Anima R. Sediment chronology in San Francisco Bay, California, defined by ^{210}Pb, ^{234}Th, ^{137}Cs, and ^{239,240}Pu. Mar Chem 1999; 64(1):7-27.

Giffin D, Corbett DR. Evaluation of sediment dynamics in coastal systems via short-lived radioisotopes. J Mar Syst 2003; 42(3):83-96.

Green MO, Coco G. Review of wave-driven sediment resuspension and transport in estuaries. Rev Geophys 2014; 52(1):77-117.

IAEA. Sediment distribution coefficients and concentration factors for biota in the marine environment. Technical Reports Series No. 422. International Atomic Energy Agency 2004.

Ji ZG. Hydrodynamics and Water Quality: Modelling Rivers, Lakes and Estuaries. Wiley 2008.

Kelderman P. Sediment pollution, transport, and abatement measures in the city canals of delft, The Netherlands. Water, Air, & Soil Pollution 2012; 223(7):4627-4645.

Kelderman P, Ang'weya R, De Rozari P, Vijverberg T. Sediment characteristics and wind-induced sediment dynamics in shallow Lake Markermeer, the Netherlands. Aquat Sci 2012; 74(2):301-313.

Khaba L, Griffiths JA. Calculation of reservoir capacity loss due to sediment deposition in the Muela reservoir, Northern Lesotho. Int Soil Water Conserv Res 2017; 5:130-140.

Kolker AS, Miner MD, Weathers HD. Depositional dynamics in a river diversion receiving basin: The case of the West Bay Mississippi River Diversion. Estuar Coast Shelf Sci 2012; 106:1-12.

Lepland A, Andersen TJ, Lepland A, Arp HPH, Alve E, Breedveld GD, Rindby A. Sedimentation and chronology of heavy metal pollution in Oslo Harbour, Norway. Mar Pollut Bull 2010; 60(9):1512-1522.

Leys V, Mulligan RP. Modelling coastal sediment transport for harbour planning: selected case studies. INTECH 2011.

Lu X. A note on removal of the compaction effect for the ^{210}Pb method. Appl Radiat Isot 2007; 65(1):142-146.

Mullenbach B, Nittrouer C, Puig P, Orange D. Sediment deposition in a modern submarine canyon: Eel Canyon, northern California. Mar Geol 2004; 211(1):101-119.

Nyarko E, Klubi E, Laissaoui A, Benmansour M. Estimating recent sedimentation rates using lead-210 in tropical estuarine systems: case study of Volta and Pra estuaries in Ghana, West Africa. J Oceanogr Mar Res 2016; 4(1):141-145

Olsen CR, Larsen IL, Lowry PD, Cutshall NH, Todd JF, Wong GTF, Casey WH. Atmospheric fluxes and marsh-soil inventories of ^{7}Be and ^{210}Pb. J Geophys Res 1985; 90(D6):10487-10495.

Ortt RA, Kerhin RT, Wells D, Cornwell J. Bathymetric survey and sedimentation analysis of Loch Raven and Prettyboy reservoirs. Coastal and Estuarine Geology File Report 2000;99-4.

Palinkas C, Nittrouer C, Wheatcroft R, Langone L. The use of [7]Be to identify event and seasonal sedimentation near the Po River delta, Adriatic Sea. Mar Geol 2005; 222:95-112.

Periáñez R, Abril JM. A numerical modeling study on oceanographic conditions in the former Gulf of Tartessos (SW Iberia): Tides and tsunami propagation. J Mar Syst 2014; 139:68-78.

Pfitzner J, Brunskill G, Zagorskis I. [137]Cs and excess ^{210}Pb deposition patterns in estuarine and marine sediment in the central region of the Great Barrier Reef Lagoon, north-eastern Australia. J Environ Radioact 2004; 76(1):81-102.

Pham MK, Povinec PP, Nies H, Betti M. Dry and wet deposition of [7]Be, ^{210}Pb and ^{137}Cs in Monaco air during 1998-2010: Seasonal variations of deposition fluxes. J Environ Radioact 2013; 120:45-57.

Qu W, Kelderman P. Heavy metal contents in the Delft canal sediments and suspended solids of the River Rhine: multivariate analysis for source tracing. Chemosphere 2001; 45(6):919-925.

Robbins JA, Holmes C, Halley R, Bothner M, Shinn E, Graney J, Keeler G, tenBrink M, Orlandini KA, Rudnick D. Time-averaged fluxes of lead and fallout radionuclides to sediments in Florida Bay. J Geophys Res 2000; 105(C12):28805-28821.

Robbins JA. Geochemical and geophysical applications of radioactive lead isotopes. In: Nriagu JO (Ed.), Biogeochemistry of Lead. Elsevier (1978).

Rose LE, Kuehl SA, 2010. Recent sedimentation patterns and facies distribution on the Poverty Shelf, New Zealand. Mar Geol 2010; 270:160-174.

Schipper C, Rietjens I, Burgess R, Murk A. Application of bioassays in toxicological hazard, risk and impact assessments of dredged sediments. Mar Pollut Bull 2010; 60(11): 2026-2042.

Schmidt S, Jouanneau J-M, Weber O, Lecroart P, Radakovitch O, Gilbert F, Jézéquel D. Sedimentary processes in the Thau Lagoon (France): from seasonal to century time scales. Estuar Coast Shelf Sci 2007a; 72(3):534-542.

Schmidt S, Gonzalez J-L, Lecroart P, Tronczyński J, Billy I, Jouanneau J-M. Bioturbation at the water-sediment interface of the Thau Lagoon: impact of shellfish farming. Aquat Living Resour 2007b; 20(2):163-169.

Sharma P, Gardner L, Moore W, Bollinger A. Sedimentation and bioturbation in a salt marsh as revealed by ^{210}Pb, ^{137}Cs, and [7]Be studies. Limnol Oceanogr 1987; 32(2):313-326.

Smith J, Lee K, Gobeil C, Macdonald R. Natural rates of sediment containment of PAH, PCB and metal inventories in Sydney Harbour, Nova Scotia. Sci Total Environ 2009; 407(17):4858-4869.

Smith JN. Why should we believe 210 Pb sediment geochronologies? J Environ Radioact 2001; 55(2):121-123.

Sommerfield CK, Nittrouer CA, Alexander CR. ^{7}Be as a tracer of flood sedimentation on the northern California continental margin. Conti Shelf Res 1999; 19:335-361.

Syvitski JP, Vörösmarty CJ, Kettner AJ, Green P. Impact of humans on the flux of terrestrial sediment to the global coastal ocean. Science 2005; 308(5720):376-380.

Tang CW-y, Ip CC-m, Zhang G, Shin PK, Qian P-y, Li X-d. The spatial and temporal distribution of heavy metals in sediments of Victoria Harbour, Hong Kong. Mar Pollut Bull 2008; 57(6):816-825.

Tanner PA, Leong LS, Pan SM. Contamination of heavy metals in marine sediment cores from Victoria Harbour, Hong Kong. Mar Pollut Bull 2000; 40(9):769-779.

Taylor A, Blake WH, Smith HG, Mabit L, Keith-Roach MJ. Assumptions and challenges in the use of fallout beryllium-7 as a soil and sediment tracer in river basins. Earth Sci Rev 2013; 126:85-95.

Thomas S, Ridd P. Review of methods to measure short time scale sediment accumulation. Mar Geol 2014; 207:95-114.

Van Rijn LC. Estuarine and coastal sedimentation problems. Int J Sediment Res 2005; 20(1):39-51.

Yeager K, Santschi P, Phillips J, Herbert B. Suspended sediment sources and tributary effects in the lower reaches of a coastal plain stream as indicated by radionuclides, Loco Bayou, Texas. Environ Geol 2005; 47(3):382-395.

Yusoff AH, Zulkifli SZ, Ismail A, Mohamed CAR. Vertical trend of trace metals deposition in sediment core off Tanjung Pelepas Harbour, Malaysia. Procedia Environ Sci 2015; 30:211-216.

Chapter 8

Settling Fluxes and Ecotoxicological Risk Assessment of Fine Sedimentary Metals in Tema Harbour (Ghana)

A modified version of this chapter was published as: Botwe, B.O., Nyarko, E., & Lens, P.N.L. (2018). Settling fluxes and ecotoxicological risk assessment of fine sedimentary metals in Tema Harbour (Ghana). *Marine Pollution Bulletin, 126*, 119-129.

Abstract

Sediment traps were deployed in the Tema Harbour to estimate the settling fluxes of silt-clay particles and associated metals (Fe, Mn, Pb, Cr, Cu, Zn, Ni, Hg, Sn and As) and characterise their potential ecotoxicological risks. The mean daily settling fluxes of the silt-clay particles and associated metals ranged from 42.7 to 85.0 g m^{-2} d^{-1} and 1.3 x 10^{-2} to 49.4 mg m^{-2} d^{-1}, respectively, and were characterised by large fluctuations at each station. The silt-clay and metal fluxes strongly correlated, indicating the important role of the silt-clay particles in metal transport and distribution in the harbour. Geochemical indices indicated anthropogenic influences on the harbour as the Pb, Cr, Zn, Hg, Sn and As content in the settling silt-clay particles exceeded average crustal concentrations. Sediment quality guidelines indicated these metals pose appreciable ecotoxicological risks, particularly As. Increasing temporal trends in As necessitates increased pollution control efforts at the harbour.

8.1. Introduction

Many West African coastal states are tapping into the blue economy by the construction of new harbours along their coasts and expansion of existing ones. These harbours are often sited close to industrial, urban and agricultural areas that serve as important sources of chemical contamination, aside contamination from shipping activities within the harbour itself (Petrosillo et al., 2009; Lepland et al., 2010; Mestres et al., 2010; Schipper et al., 2010; Botwe et al., 2017a). Chemical contamination in harbours poses a threat to their sustainability as it can result in adverse impacts such as hindrance to harbour activities, alteration of ecological habitats, death of marine organisms resulting in loss of biodiversity, decline in productivity, fishing restrictions, dietary restrictions on seafood (Lepland et al., 2010), increased incidence of human diseases and high costs of maintenance dredging (Botwe et al., 2017b). Although environmental issues and sustainability are becoming an integral part of harbour development, operation and management (Mestres et al., 2010), effective regulation of contaminant discharges into harbours from diffuse sources still remains a challenge (Petrosillo et al., 2009), particularly for developing countries. Thus, continuous monitoring and assessment of chemical contamination in harbours is crucial for their management (Pozza et al., 2004; Botwe et al., 2017c).

Metals are among the widespread environmental contaminants due to their wide societal use and applications. Metals such as Pb, Cr, Cu, Zn, Ni, As, Sn and Hg are not only hazardous to humans and non-human biota, but are also persistent and bioaccumulative (Casado-Martínez et al., 2006; Nyarko et al., 2011; Botwe et al., 2017d). Therefore, their contamination deserves great attention. In aquatic environments, metals are scavenged and accumulated by fine sedimentary materials, particularly silt-clay (< 63 µm) particles, as a result of their high sorptive capacity (Horowitz, 1985; Horowitz and Elrick, 1987; Horowitz et al., 1989; Lepland et al., 2010; Hamzeh et al., 2014). Thus, the silt-clay particles play a major role in the fate of metals in aquatic ecosystems, including their fluxes, deposition, burial and resuspension (Hostache et al., 2014; Botwe et al., 2017b). Therefore, in order to adequately address the problem of metal contamination in aquatic ecosystems, the fine sedimentary materials should be the prime materials for investigation (Horowitz, 1985). Moreover, information on the fluxes of fine sedimentary materials and associated metals can provide an insight into recent anthropogenic influences on the ecosystem under study and hence, the effectiveness of metal pollution control measures.

Sediment traps, which passively collect and accumulate settling particulates, have proven useful for the investigation of particulate matter fluxes in lacustrine (Hakanson, 2006; de Vicente et al., 2010; Kelderman et al., 2012) and marine (Buesseler et al., 2007; Santos-Echeandía et al., 2012; Liu et al., 2014; Szmytkiewicz and Zalewska, 2014; Helali et al., 2016) ecosystems. To enhance trap efficiency and minimise loss of trapped material as a result of current-induced resuspension, sediment traps with aspect ratios (i.e. height/diameter) > 5 are recommended (Bloesch & Burns, 1980; Kelderman et al., 2012; de Vicente et al., 2010). Retrieving sediment traps after 2 weeks of deployment is also recommended as it allows accumulation of sufficient particulate matter, while minimising organic matter decomposition (Hakanson and Jansson, 1983; Buesseler et al., 2007; Kelderman et al., 2012).

The coastal marine Tema Harbour in Ghana is one of the large harbours in West Africa and has a long history of urbanisation and industrialisation within its catchments (Botwe et al, 2017b). Due to a lack of regulations on industrial discharges or weak enforcement in Ghana, the Tema Harbour may act as a repository for contaminants from point and diffuse sources. In order to understand and minimise anthropogenic impact on the harbour, it is essential to characterise contaminant fluxes, sources and associated ecotoxicological risks (Pozza et al., 2004). However, existing literature does not provide information on metal fluxes in the harbour. The objectives of this study, therefore, were to (1) estimate settling fluxes of silt-clay particles and associated metals (Mn, Pb, Cr, Cu, Zn, Ni, Hg, Sn and As) in the Tema Harbour,

(2) quantify the extent of metal contamination in the sediments and (3) characterise the potential ecotoxicological risks.

8.2. Materials and methods

8.2.1. Study area

The Tema Harbour is situated along the Ghana sea coast at Tema, the industrial city of Ghana (Botwe et al., 2017c). It was constructed in 1960 to support industrial growth in Ghana and is partitioned into a Main Harbour, an Inner Fishing Harbour, an Outer Fishing Harbour and a Canoe Basin (Botwe et al., 2017c) with a water area of 1.7 km^2, an average tidal range of 1.6 m and tidal currents in the range of 0.1-0.5 ms-1 (Botwe et al. 2017b). The Main Harbour is designated for shipping activities and has a 240 m wide entrance, a breakwater of 4850 m length and water depths in the range of 7.5 to 11.4 m (Botwe et al., 2017b, e). The Fishing Harbours and the Canoe Basin provide landing and fuel storage facilities that sustain a productive fishery (Botwe et al., 2017a, d). Notable industrial activities in Tema are crude oil refining, alumina smelting, paint, steel and cement production (Botwe et al., 2017c). The main land uses in the Tema metropolis are residential, agricultural (largely subsistence), industrial and commercial. Industrial effluents, domestic wastes and wastewater from the industrial areas and other parts of the Tema township may eventually reach the Tema Harbour via drains, surface runoff and alongshore transport.

8.2.2. Sediment sampling and analyses

The sediment samples were collected as part of a broader study on sedimentation conditions in the Tema Harbour, in which the sampling methodology has been described (Botwe et al. 2017b). In that study, settling fluxes of bulk sediment in the water column and sediment accumulation rates in the Tema Harbour were quantified, while the present study quantified the settling fluxes of only the silt-clay particles and their associated metals as they represent the most bioavailable and ecologically relevant sediment grain-sizes (Bat and Raffaelli, 1998; Bat, 2005). Briefly, cylindrical sediment traps made of polyvinyl chloride (PVC) with aspect ratios of 6.0 were deployed at five stations (I, II, III, IV and V) along a transect in a less busy area within the Main Harbour (see Fig. 1 in Botwe et al., 2017b), where the water depths ranged from 9 to 11 m. At each station, a vertical array of two sediment traps, positioned 0.6 m apart in an alternating fashion, were deployed to collect settling particles at two water depths of approximately 1.8 m (top trap) and 0.6 m (bottom trap) from the seabed. The sediment traps were retrieved every two weeks over a period of 12 weeks (23 May - 8 August, 2015). The sediment traps were capped prior to their retrieval in order to keep the trapped materials intact. A total of 56 sediment trap samples were collected over the study period. Following retrieval, the sediment traps with their contents were kept in upright positions in a rack and transported to the Department of Marine and Fisheries Sciences laboratory (University of Ghana) for further processing and analysis.

In the laboratory, the trapped sediment samples were separated into silt-clay (< 63 μm) and sand (> 63 μm) particles by wet-sieving over a 63 μm stainless steel mesh and then oven-dried at 50°C till constant weight (Wang et al., 2001; Botwe et al., 2017c). Only the silt-clay particles were processed for metal, total organic carbon (TOC) and total nitrogen (TN) analyses. In this study, metals were analysed in the silt-clay particles in order to minimise the potential effect of grain size on the metal distribution (Horowitz, 1985). Prior to the analyses of metals, TOC and TN, the silt-clay samples were homogenised using a Fritsch Pulveriser (Botwe et al., 2017d). For the metal analysis, 2.5 ml of 65% HNO_3 was added to about 0.25 g portions of the sediment samples in Teflon tubes, placed in a stainless steel bomb, and digested on a hot plate at 120°C for 4 h following the Canadian National Laboratory for Environmental Testing (1994) protocol. Each digested sediment sample was quantitatively transferred into a polypropylene tube using deionised water to obtain 50 ml final solution and subsequently assayed for metal concentrations by Atomic Absorption Spectrometry (Varian AA 240FS) with deuterium background correction. Mn, Pb, Cr, Cu, Zn, Ni and Fe were analysed using the Flame Technique, Sn and As by the Graphite Furnace Technique, and Hg by the Cold Vapour Technique. The detection limits for the analysed metals were: 0.10 mg kg-1 dw for Hg and As; 0.50 mg kg-1 dw for Mn, Pb, Cr, Zn, Ni, Sn and Fe; and 1.0 mg kg-1 dw for Cu. The TOC and TN contents of the sediments were determined on 0.5 g of dry sediment samples using the Walkley-Black wet oxidation and the Kjeldahl methods, respectively, following Botwe et al. (2017c).

8.2.3. Quality control/assurance

Prior to the metal analyses, auto calibration of the AAS was made by aspirating a bulk standard (Multi-elemental standard solution for ICP, Fluka Analytical, Switzerland). Procedural blanks and a Certified Reference Material for marine sediments (IAEA-158) were processed in a similar way as the field samples and analysed to check the quality of all results.

Sediment samples were analysed in duplicates while reference materials were analysed in triplicates. Chemicals, solvents and reagents used were of trace metal analysis grade (from Sigma-Aldrich, U.S.A). All containers were thoroughly washed with detergent, soaked in a 10% HNO_3 solution overnight, rinsed with deionised water and dried in an oven prior to their use. Sediment traps were also conditioned with dilute nitric acid and rinsed with distilled water prior to their use. Metal recoveries were in the range of 63-95.5 %, with relative standard deviations (RSD) of <5% (not shown).

8.2.4. Data analysis

One-way analysis of variance (ANOVA) and the Holm-Sidak multiple comparison test were performed to assess the spatial and temporal variations in the accumulated dry mass of silt-clay and their associated metal and TOC contents using the statistical software SigmaPlot (version 11.0). When the normality test failed, Kruskal-Wallis one-way ANOVA on ranks was performed instead of the Holm-Sidak multiple comparison test. Two-tailed Pearson correlations were conducted to examine linear relationships among the measured metals, accumulated dry mass of silt-clay and TOC concentrations using SPSS (version 16.0). Descriptive statistics such as means, standard deviations and standard errors were computed in Microsoft Excel 2007 at the 95% confidence level.

The settling fluxes of silt-clay (F_s), with units of g m^{-2} d^{-1}, were estimated using Eq. 1 (Botwe et al., 2017b):

$$F_s = (M)/(A \times D) \tag{1}$$

where M is the dry accumulated mass of silt-clay (kg) in the sediment trap, A is the cross-sectional area of the sediment trap (m^2) and D is the duration of trap deployment (d).

The settling fluxes of silt-clay associated metals (F_m, with units of mg m^{-2} d^{-1}) were estimated using Eq. 2:

$$F_m = F_s \times C_m \tag{2}$$

where C_m is the metal concentration in the silt-clay particles (mg g^{-1} d.w).

Metal contamination in sediments may originate from lithogenic and anthropogenic sources (Clark, 2001). To delineate lithogenic from anthropogenic metal contamination in the Tema Harbour sediments, metal enrichments in the sediments were assessed against average crustal metal concentrations (Taylor, 1964) by deriving the metal enrichment factors (EFs) using Fe as a normaliser according to Eq. (3) (Addo et al., 2011; Yilgor et al., 2012):

$$EF = [M/Fe]_{Sample}/[M/Fe]_{Crust} \qquad\qquad (3)$$

where $[M/Fe]_{Sample}$ is the metal-iron ratio in the Tema Harbour sediments and $[M/Fe]_{Crust}$ is the metal-iron ratio based on the average crustal concentrations reported by Taylor (1964) as proxies for their background concentrations in the Tema Harbour and were: Fe = 56,300; Mn = 950; Pb = 12.5; Cr = 100; Cu = 55; Zn = 70; Ni = 75; Hg = 0.08; Sn = 2; As = 1.8 mg.kg^{-1}. By this approach, the progression of metal enrichment is as follows: deficient (EF ≤ 1), minor enrichment (1 < EF ≤ 3), moderate enrichment (3 < EF ≤ 5), moderately severe enrichment (5 < EF ≤ 10), severe enrichment (10 < EF ≤ 25), very severe enrichment (25 < EF ≤ 50), and extremely severe enrichment (EF > 50).

Quantitative assessment of metal contamination in the Tema Harbour sediments was conducted by deriving metal geo-accumulation indices (Igeo) (Muller, 1969), following Botwe et al. (2017d) according to Eq. (4):

$$Igeo = Log_2 [Cn/(1.5 \times Bn)] \qquad\qquad (4)$$

where Cn is the metal concentration in the Tema Harbour sediment and Bn is the corresponding average crustal concentration (Taylor, 1964) taken as a proxy for the metal background concentration (Botwe et al., 2017b). The Igeo quantifies the progression of metal contamination in sediments as follows (Muller, 1969): uncontaminated (Igeo < 0), uncontaminated to moderately contaminated (0 ≤ Igeo < 1), moderately contaminated (1 ≤ Igeo < 2), moderately to highly contaminated (2 ≤ Igeo < 3), heavily contaminated (3 ≤ Igeo < 4), highly to very highly contaminated (4 ≤ Igeo < 5), very heavily contaminated (Igeo ≥ 5).

8.3. Results

8.3.1. Metals, TOC and TN concentrations in settling silt-clay particles in the Tema Harbour

Table 8.1 presents the accumulated dry masses of sediment fractions in the sediment traps and silt-clay associated metal, TOC and TN contents at the different sampling stations and for the sampling periods. The accumulated sediments in the sediment traps were mainly composed of sand with dry masses ranging from 7.8 to 412 g, representing 67-99 % of the dry bulk sediment mass accumulated in the traps, while the dry masses of the silt-clay particles were in the range of 1.9-38.5 g, representing 1-33 % of the dry bulk sediment mass accumulated in the traps. The silt-clay particles contained varying concentrations of metals with Fe occurring in relatively higher concentrations. Among the other metals, Mn (260-1083 mg.kg^{-1} dw) and As (146-1470 mg.kg^{-1} dw) were present in relatively high concentrations. The Hg concentrations were relatively low and ranged from <0.1 to 3.0 mg.kg^{-1} dw, while those of Pb, Cr, Cu, Zn, Ni and Sn were in the ranges of 20.3-229, 22.5-373, 15.8-119, 94.6-354, 18.4-66.1 and 6.2-49.2 mg.kg^{-1} dw, respectively. The TOC content of the accumulated silt-clay particles ranged from 2.1 to 13.3%, while the TN content ranged from 0.42 to 0.98% with corresponding TOC/TN ratios in the range of 3.3-27.7 (Table 8.1).

Table 8.2 presents the correlation matrix of the metal concentrations with the TOC content. The results show a strong direct correlation between Cu and Zn concentrations ($r = 0.76$, p <0.01). Both Cu and Zn exhibited statistically significant moderate correlations with Pb and Fe. Cr exhibited statistically significant moderate correlations with Cu ($r = 0.49$, p <0.01) and Sn ($r = 0.45$, p <0.01), but weak correlations with Fe ($r = 0.37$, p <0.01) and Ni ($r = 0.28$, p <0.05). Hg and Sn were also correlated ($r = 0.44$, p <0.01), while Mn exhibited weak but statistically significant inverse correlations with Pb ($r = -0.28$, p <0.05) and As ($r = -0.27$, p <0.05). There were no statistically significant correlations between the TOC and the measured metals.

Table 8.1 Accumulated sediment fractions in sediment traps deployed at different sampling stations (see Fig. 7.1) and periods in Tema Harbour (Ghana) and the silt-clay particles associated metal, TOC and TN contents

Sampling period	Sampling station	Trap position	Sediment fractions		Silt-clay particles associated metal concentrations (mg kg^{-1} dw)										TOC (%)	TN (%)	TOC/TN
			Silt-clay (g dw)	Sand (g dw)	Mn	Pb	Cr	Cu	Zn	Ni	Hg	Sn	As	Fe			
Week 2	I	Top	5.1	24.2	418	65.0	189	63.6	244	34.4	0.40	24.0	374	35740	3.6	0.64	5.6
		Bottom	2.7	26.9	611	47.3	217	64.9	251	54.8	0.47	35.5	147	40840	9.6	0.62	15.5
	II	Top	4.1	36.7	530	55.4	187	59.7	236	23.2	0.43	22.6	192	51940	4.0	0.64	6.3
		Bottom	3.1	42.6	607	75.3	210	62.5	230	27.0	0.30	22.8	226	40740	3.2	0.48	6.7
	III	Top	2.0	120.1	289	229	231	67.7	271	22.5	0.41	31.2	440	49240	2.4	0.64	3.8
		Bottom	2.3	181.5	282	26.7	132	21.3	210	18.4	0.12	25.5	259	37440	2.3	0.52	4.4
	IV	Top	4.5	119.8	592	28.3	167	49.3	193	23.0	0.23	16.4	148	40590	9.2	0.42	21.9
		Bottom	3.6	95.4	510	43.9	200	52.5	206	31.5	0.50	25.8	336	52840	8.4	0.58	14.5
	V	Top	7.2	268	770	21.0	162	50.4	195	38.6	0.42	24.2	614	56300	4.4	0.7	6.3
		Bottom	2.0	214.8	1183	20.3	175	40.8	168	22.5	0.36	20.4	146	36380	9.2	0.42	21.9
Week 4	I	Bottom	4.9	45.1	616	65.3	204	54.5	193	54.7	0.24	31.8	454	41770	5.0	0.50	10.0
	II	Top	5.2	89.6	640	54.9	184	50.3	196	54.4	0.20	21.5	403	39170	2.1	0.53	4.0
		Bottom	4.7	69.1	678	36.3	32.5	38.2	187	63.7	0.30	8.5	443	37610	4.0	0.45	8.9
	III	Top	9.8	113.5	548	26.4	26.5	29.6	175	44.5	0.17	6.2	403	40330	3.6	0.86	4.2
		Bottom	4.9	249.2	523	21.6	41.3	34.5	186	47.8	0.10	6.8	522	43230	5.2	0.5	10.4
	IV	Top	9.3	69.2	616	33.2	199	45.4	195	55.7	0.30	14.1	493	49880	8.0	0.81	9.9
		Bottom	10.5	30.2	517	50.3	287	49.6	185	56.6	0.16	20.8	758	47710	3.6	0.47	7.7
	V	Top	9.9	88.0	553	41.6	61.5	38.5	188	52.1	0.26	32.9	352	38530	5.2	0.45	11.6
		Bottom	3.5	76.3	709	22.6	196	43.9	189	52.8	0.30	35.6	373	43440	8.4	0.5	16.8
Week 6	I	Top	4.1	11.8	538	45.6	225	65.8	208	66.1	0.42	32.3	572	63690	4.4	0.56	8.5
		Bottom	4.4	19.9	563	45.5	231	57.7	198	62.7	0.27	29.8	656	54130	6.4	0.44	14.5
	II	Top	20.6	68.6	598	42.2	202	46.5	172	53.2	0.51	25.0	606	35860	4.0	0.47	8.5
		Bottom	1.9	7.8	486	41.1	231	55.5	190	61.7	0.25	22.6	427	42690	6.2	0.58	10.7
	III	Top	3.6	84.2	483	34.1	204	48.5	196	37.1	0.20	35.0	508	40530	5.8	0.45	12.9
		Bottom	6.2	24.3	284	49.6	146	36.5	217	39.2	0.38	24.7	480	39330	2.4	0.72	3.3
	IV	Top	11.0	77.9	541	44.4	172	37.0	150	42.5	1.33	39.9	539	41670	7.9	0.98	8.1
		Bottom	38.5	139.6	657	31.7	195	43.4	171	47.2	3.00	47.3	612	50020	6.0	0.53	11.3

Table 8.1 continued

Sampling period	Sampling Station	Trap position	Sediment fractions		Silt-clay particles associated metal concentrations (mg kg⁻¹ dw)										TOC (%)	TN (%)	TOC/TN
			Silt-clay (g dw)	Sand (g dw)	Mn	Pb	Cr	Cu	Zn	Ni	Hg	Sn	As	Fe			
	V	Top	5.7	34.2	510	50.2	236	36.3	168	53.1	0.25	34.7	683	35180	7.2	0.82	8.8
		Bottom	5.8	42.4	511	41.3	212	26.1	171	39.7	0.10	49.2	491	38010	4.4	0.67	6.6
Week 8	I	Bottom	3.8	33.1	402	31.8	22.5	21.9	95	22.4	0.10	22.8	378	37040	7.8	0.55	14.2
	II	Top	3.1	64.8	563	47.6	34.3	15.8	154	19.4	0.10	24.4	382	40460	6.4	0.58	11.0
	III	Top	5.4	350	422	47.6	61.0	33.9	157	26.9	0.18	24.5	551	37860	3.2	0.57	5.6
		Bottom	7.7	153	406	46.6	136	43.9	164	35.4	0.23	22.2	603	39850	4.4	0.54	8.1
	IV	Top	7.1	184	426	51.2	215	40.4	164	27.5	0.28	24.8	278	36080	7.2	0.5	14.4
		Bottom	8.6	26.9	679	31.8	179	31.9	156	31.0	0.30	27.7	244	34900	4.7	0.62	7.6
	V	Top	2.1	32.7	637	29.8	181	38.2	153	46.5	0.48	20.4	617	48980	7.2	0.54	13.3
		Bottom	5.4	47.8	537	25.3	184	40.8	169	45.6	0.28	30.6	569	44950	4.4	0.56	7.9
Week 10	I	Top	4.5	118.4	368	31.6	138	48.6	162	34.9	0.21	22.5	520	35720	4.4	0.5	8.8
		Bottom	4.6	155.2	406	50.2	373	62.2	206	58.4	0.29	35.9	1137	57960	6.5	0.53	12.3
	II	Top	6.3	112.2	597	80.2	270	70.1	226	44.3	0.53	38.5	775	65920	5.2	0.62	8.4
		Bottom	3.7	38.5	260	44.8	245	49.7	149	38.3	0.22	16.7	615	37600	4.0	0.42	9.5
	III	Top	5.3	111	409	47.4	200	50.9	174	35.4	0.10	23.6	342	37100	8.4	0.44	19.1
		Bottom	2.7	49.7	402	32.4	210	47.7	171	36.9	0.30	23.5	399	26240	8.8	0.95	9.3
	IV	Top	3.6	66.8	309	48.6	173	32.4	120	27.1	0.54	20.7	628	39360	3.6	0.45	8.0
		Bottom	2.3	28.7	438	41.5	233	47.6	157	44.1	0.81	28.9	573	53920	5.8	0.73	7.9
	V	Top	4.8	49.2	557	35.0	180	119.4	354	28.6	0.20	31.5	1051	49340	6.0	0.48	12.5
		Bottom	3.8	56.1	460	44.1	232	42.4	156	39.6	0.26	33.9	591	56070	6.3	0.45	14.0
Week 12	I	Bottom	8.7	72.4	358	51.1	228	64.1	212	35.6	1.12	34.4	888	34550	4.8	0.42	11.4
	II	Top	2.4	20.7	400	34.6	179	46.7	136	27.8	0.41	25.9	1054	38010	4.0	0.53	7.5
		Bottom	7.5	28.1	434	41.3	195	50.4	186	32.2	0.71	23.6	1137	40050	6.7	0.45	14.9
	III	Top	5.5	11.2	294	42.0	187	45.5	144	31.3	0.44	22.3	1467	33290	7.9	0.67	11.8
		Bottom	5.8	68.8	368	32.7	106	47.8	166	33.2	0.81	31.9	1470	39810	5.2	0.62	8.4
	IV	Top	8.2	412	469	43.4	267	55.3	177	41.4	0.21	29.0	1045	49920	4.7	0.49	9.6
		Bottom	4.2	31.5	786	37.6	270	53.8	173	45.1	0.40	26.5	1161	44710	2.8	0.53	5.3
	V	Top	4.1	30.8	494	38.2	276	46.8	152	45.4	0.93	31.4	426	44710	13.3	0.48	27.7
		Bottom	5.3	33.6	580	43.1	226	57.9	155	42.8	0.28	26.4	434	34970	13.3	0.67	19.9

Table 8.2 Correlation matrix of metal and TOC concentrations in settling silt-clay particles in the Tema Harbour, Ghana ($n = 56$)

	Mn	Pb	Cr	Cu	Zn	Ni	Hg	Sn	As	Fe
Pb	-0.28*									
Cr	-0.01	0.23								
Cu	0.04	0.30*	0.49**							
Zn	0.07	0.38**	0.20	0.76**						
Ni	0.23	-0.14	0.28*	0.14	0.05					
Hg	0.07	-0.01	0.14	0.05	-0.06	0.07				
Sn	-0.01	0.16	0.45**	0.19	0.14	0.10	0.44**			
As	-0.27*	-0.05	0.24	0.27*	-0.05	0.09	0.17	0.17		
Fe	0.16	0.16	0.37**	0.38**	0.32*	0.32*	0.16	0.25	0.16	
TOC	0.21	-0.26	0.18	0.04	-0.19	0.05	0.12	0.12	-0.14	-0.07

*Correlation is significant at the 0.05 level (2-tailed)

**Correlation is significant at the 0.01 level (2-tailed)

8.3.2. Distributions and fluxes of silt-clay particles associated metals in the Tema Harbour

At all the sampling stations, t-tests revealed no significant differences ($p > 0.05$) between the mean silt-clay particles associated metal concentrations in the top and bottom traps. Similarly, the mean TOC content in the silt-clay did not differ significantly ($p > 0.05$) between the top and bottom traps across the sampling stations. Therefore, the mean metal concentrations for the top and bottom traps at each station were determined to assess the temporal and spatial distributions of the metals, which are presented in Figs. 8.1 and 8.2, respectively. One-way ANOVA with pairwise multiple comparison procedures (Holm-Sidak method) and Kruskal-Wallis one way ANOVA on ranks with pairwise multiple comparison procedures (Dunn's Method) revealed statistically significant temporal variations in the concentrations of Ni (p <0.001), Cu ($p = 0.010$), Zn (p <0.001), Sn ($p = 0.034$) and As (p <0.001). Relatively higher concentrations of Ni occurred in weeks 4 and 6 (Fig. 8.1f), the lowest concentrations of Cu (Fig. 8.1d) and Zn (Fig. 8.1e) occurred in week 8, the highest concentration of Sn was observed during week 6 (Fig. 8.1h), while the concentrations of As increased over the period with a decrease during the eighth week (Fig. 8.1i).

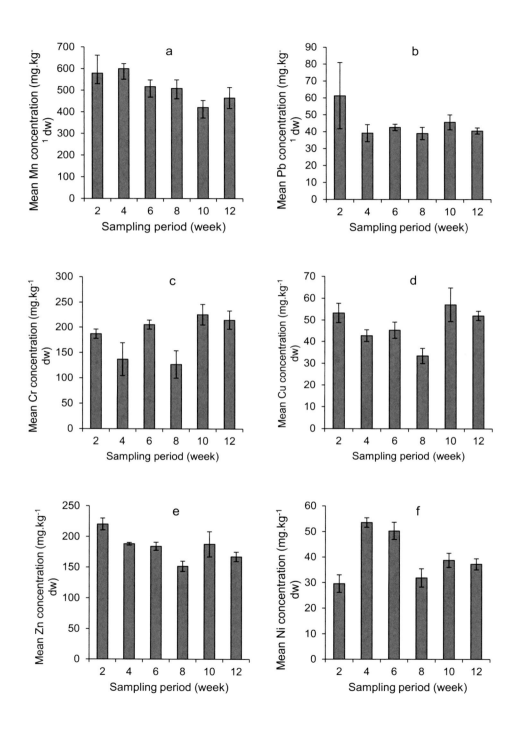

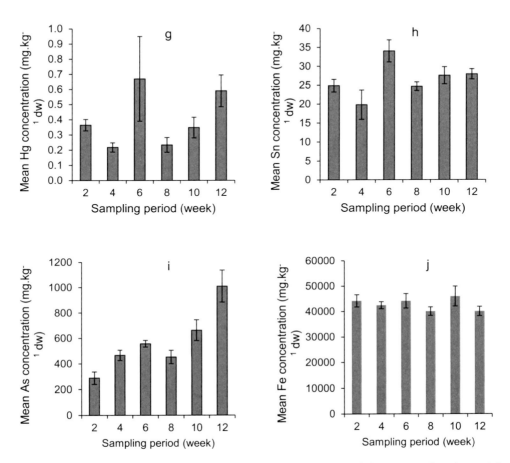

Fig. 8.1 Temporal distribution of settling silt-clay particles associated metal concentrations (mean ± standard error) in the Tema Harbour (Ghana): (a) Mn, (b) Pb, (c) Cr, (d) Cu, (e) Zn, (f) Ni, (g) Hg, (h) Sn, (i) As and (j) Fe.

The spatial trends in the metal concentrations in the settling silt-clay particles in the Tema Harbour are shown in Fig. 8.2. Among the analysed metals, Kruskal-Wallis one-way ANOVA on ranks with pairwise multiple comparison procedures (Dunn's Method) revealed statistically significant spatial variations in the concentrations of Mn ($p = 0.003$), Pb ($p = 0.041$) and Cu ($p = 0.032$). Relatively lower concentrations of Mn (Fig. 8.2a) and Cu (Fig. 8.2d) were observed at station III, while relatively lower concentrations of Pb were observed at station V (Fig. 8.2b).

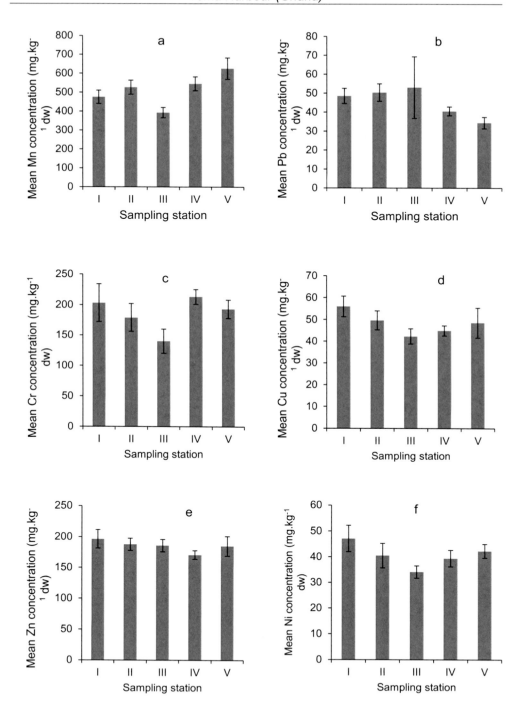

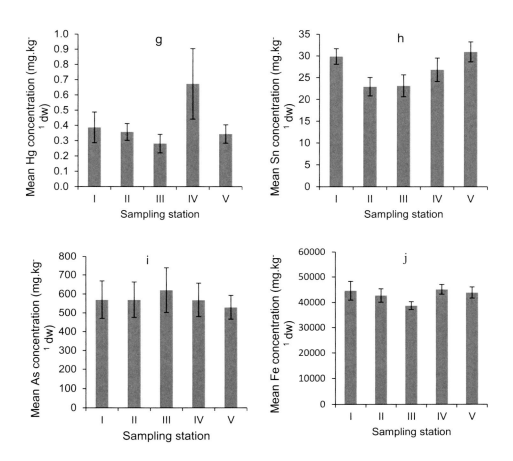

Fig. 8.2 Spatial distribution of settling silt-clay particles associated metal concentrations (mean ± standard error) in the Tema Harbour (Ghana): (a) Mn, (b) Pb, (c) Cr, (d) Cu, (e) Zn, (f) Ni, (g) Hg, (h) Sn, (i) As and (j) Fe. See sampling stations A-E in Fig. 7.1.

The estimated mean daily settling fluxes of silt-clay and associated metals at the five sampling stations in the Tema Harbour are presented in Table 8.3. The mean daily settling fluxes of silt-clay across the sampling stations ranged from 42.7 (± 5.1) to 85.0 (± 31) $g.m^{-2}.d^{-1}$, while the mean fluxes of associated metals across the sampling stations ranged from 15.8 (± 3.9) to 49.4 (± 20.4) $mg.m^{-2}.d^{-1}$ for Mn, 1.6 (± 0.4) to 3.3 (± 1.0) $mg.m^{-2}.d^{-1}$ for Pb, 5.7(± 1.3) to 17.6(± 6.1) $mg.m^{-2}.d^{-1}$ for Cr, 1.8 (± 0.6) to 3.7 (± 1.4) $mg.m^{-2}.d^{-1}$ for Cu, 6.4 (± 1.9) to 14.9

Table 8.3 Estimated daily settling fluxes of silt-clay particles and associated metals (mean ± standard error) at different sampling stations over a 12-week period in the Tema Harbour (Ghana)

Sampling station	Silt-clay flux ($g.m^{-2}.d^{-1}$)	Metal flux								
		Mn*	Pb*	Cr*	Cu*	Zn*	Ni*	Hg**	Sn*	As*
[a]I	42.7 ± 5.1	15.8 ± 3.9	1.6 ± 0.5	6.8 ± 2.2	1.8 ± 0.6	6.4 ± 1.9	2.0 ± 0.4	14.9 ± 8.4	2.1 ± 0.3	18.5 ± 7.2
[b]II	50.6 ± 17.0	26.2 ± 10.2	2.4 ± 0.8	8.9 ± 3.8	2.4 ± 0.9	8.9 ± 3.1	2.0 ± 0.9	20.3 ± 10.7	1.2 ± 0.5	29.0 ± 13.4
[c]III	46.3 ± 8.7	19.0 ± 4.7	2.0 ± 0.5	5.7 ± 1.3	1.9 ± 0.3	8.3 ± 1.5	1.7 ± 0.4	13.0 ± 4.5	1.0 ± 0.2	30.2 ± 10.1
[c]IV	85.0 ± 31.0	49.4 ± 20.4	3.3 ± 1.0	17.6 ± 6.1	3.7 ± 1.4	14.9 ± 5.8	3.0 ± 0.8	114 ± 85	2.8 ± 1.5	45.2 ± 17.8
[c]V	45.1 ± 7.7	26.9 ± 4.2	1.6 ± 0.4	8.2 ± 1.3	2.2 ± 0.5	8.5 ± 1.6	2.0 ± 0.5	14.5 ± 2.9	1.4 ± 0.3	24.1 ± 4.9

*Measured in $mg.m^{-2}d^{-1}$

**Measured in $\mu g.m^{-2}d^{-1}$; [a]$n = 9$; [b]$n = 11$; [c]$n = 12$

($\pm$ 5.8) mg.m^{-2}.d^{-1} for Zn, 1.7 ($\pm$ 0.4) to 3.0 ($\pm$ 0.8) mg.m^{-2}.d^{-1} for Ni, 13.0 ($\pm$ 4.5) to 114 ($\pm$ 85) µg.m^{-2}.d^{-1} for Hg, 1.0 ($\pm$ 0.2) to 2.8 ($\pm$ 1.5) mg.m^{-2}.d^{-1} for Sn, and 18.5 ($\pm$ 7.2) to 45.2 ($\pm$ 17.8) mg.m^{-2}.d^{-1} for As. The settling fluxes of Pb, Cu, Ni and Sn were comparable, being an order of magnitude lower than those of Mn and As, but about two orders of magnitude higher than those of Hg. Relatively higher silt-clay and metal fluxes occurred at the more outer station IV although no clear spatial trend was observed.

There were strong correlations between the silt-clay fluxes and the fluxes for the different metals (*r* values ranging from 0.87 to 0.96, *p* < 0.01), while the fluxes for the different metals investigated were also strongly correlated (*r* values ranging from 0.71 to 0.99, *p* <0.01) (Table 8.4).

Table 8.4 Correlation matrix of silt-clay particle and metal fluxes in the Tema Harbour, Ghana (*n* = 56)

	Silt-clay	Mn	Pb	Cr	Cu	Zn	Ni	Hg	Sn
Mn	0.96								
Pb	0.87	0.86							
Cr	0.95	0.97	0.87						
Cu	0.94	0.96	0.93	0.95					
Zn	0.95	0.97	0.92	0.94	0.99				
Ni	0.91	0.97	0.91	0.94	0.98	0.98			
Hg	0.93	0.90	0.71	0.89	0.85	0.86	0.83		
Sn	0.95	0.98	0.82	0.97	0.94	0.95	0.95	0.94	
As	0.89	0.89	0.92	0.88	0.93	0.96	0.92	0.81	0.89

All correlations are significant at the 0.01 level (2-tailed)

8.4. Discussion

8.4.1. Metal distribution and contamination in settling silt-clay partices in the Tema Harbour

The settling sediments in the Tema Harbour were characterised by a predominance of sand, which may be attributed to its quicker removal from the water column than the silt-clay particles (Eisma, 1993; Ji, 2008). The Ni, Cu, Zn, Sn, As Mn, Pb and Cu concentrations in the silt-clay particles exhibited temporal and/or spatial variations. Generally, variations in sediment-metal concentrations are commonly attributed to variations in silt-clay, Fe and Mn oxides and hydroxides as well as organic matter content in sediments as these tend to concentrate metals (Horowitz, 1985; Horowitz and Elrick, 1987; Horowitz et al., 1989;

Hamzeh et al., 2014). The analysis of the silt-clay particles might have diminished the potential effect of TOC on sediment metal distribution, resulting in the poor correlations between TOC and metal concentrations (Table 8.2). The statistically significant Cu/Fe, Zn/Fe, Cr/Fe, Pb/Mn and As/Mn correlations indicate a potential influence of Fe oxides and hydroxides on Cu, Zn and Cr distributions as well as of Mn oxides and hydroxides on Pb and As distributions in the harbour sediments.

The calculated EFs indicated varying levels of metal enrichment in the Tema Harbour sediments, which are summarised in Table 8.5. The harbour sediments were predominantly characterised by severe As enrichment, moderate to severe Pb, Zn and Sn enrichments, deficient to minor Mn, Cr, Cu and Hg enrichments, and were deficient in Ni. The EFs of As and Sn (in 91% of the cases) were >10, indicating they were mainly derived from anthropogenic inputs (Addo et al., 2011; Yilgor et al., 2012; Botwe et al., 2017d). For Pb, Cr, Zn and Hg, the EFs were mostly between 1.5 and 10, indicating considerable anthropogenic inputs of these metals in the Tema Harbour sediments. On the contrary, the low EFs (<1.5) of Mn, Cu and Ni suggest mainly lithogenic sources of these metals in the Tema Harbour sediments (Addo et al., 2011; Yilgor et al., 2012; Botwe et al., 2017d).

The computed I_{geo} values, summarised in Table 8.5, indicated that metal enrichment in the Tema Harbour sediments had reached varying pollution levels. All the harbour sediments were potentially very heavily polluted with As, while potentially high Sn pollution and moderate Pb, Cr and Zn pollution occurred in the majority of the sediments (over 60%). The sediments exhibited a predominance of moderate to very high Hg pollution, but were potentially unpolluted with Mn and Ni. The I_{geo} values also indicated that Cu enrichment in the sediments had not reached pollution levels. These results indicate considerable anthropogenic inputs of As, Sn, Hg, Pb, Zn and Cr as compared to Mn, Cu and Ni in the Tema Harbour, which agrees well with findings from a previous study on bottom sediments in the harbour (Botwe et al., 2017d).

Anthropogenic metal inputs into the Tema Harbour may derive principally from shipping and fishing activities such as sand blasting, scraping and painting of ships, bunkering, bilge discharges, debalasting, refueling and associated oil spills (Lepland et al., 2010; Mestres et al., 2010; Nyarko et al., 2014; Botwe et al., 2017a, 2017d) as well as metal-containing wastewater and particulates discharged from oil refinery, cement production, ore smelting, automobile

Table 8.5 Metal enrichment and pollution levels and their incidence of occurrence in settling silt-clay particles in Tema Harbour (Ghana)

Enrichment level	Incidence of occurrence for measured metals (%, $n = 56$)								
	Mn	Pb	Cr	Cu	Zn	Ni	Hg	Sn	As
No enrichment	94.6	0	12.5	39.3	0	100	39.3	1.8	0
Minor enrichment	5.4	16.1	62.5	60.7	28.6	0	30.4	0.0	0
Moderate enrichment	0	46.4	25.0	0	66.1	0	26.8	73.2	0
Moderate to severe enrichment	0	35.7	0	0	5.4	0	1.8	25.0	0
Severe enrichment	0	1.8	0	0	0	0	1.8	0	100
Very severe enrichment	0	0	0	0	0	0	0	0	0
Extremely severe enrichment	0	0	0	0	0	0	0	0	0
Contamination status	Incidence of occurrence for measured metals (%, $n = 56$)								
	Mn	Pb	Cr	Cu	Zn	Ni	Hg	Sn	As
Uncontaminated	100	16.1	25.0	100	1.8	100	8.9	0	0
Uncontaminated to moderately contaminated	0	75.0	75.0	0	76.8	0	19.6	0	0
Moderately contaminated	0	7.1	0	0	21.4	0	48.2	5.4	0
Moderately to highly contaminated	0	1.8	0	0	0	0	16.1	26.8	0
Heavily contaminated	0	0	0	0	0	0	5.4	64.2	0
Highly to very highly contaminated	0	0	0	0	0	0	1.8	3.6	0
Very heavily contaminated	0	0	0	0	0	0	0	0	100

Metal enrichment levels and contamination status were based on their enrichment factors (EFs) and geo-accumulation indices (I_{geo}), respectively. For interpretation of the EF and I_{geo} values, see Section 8.2.4.

exhaust and agrochemical use within the harbour area (Yilgor et al., 2012; El-Sorogy et al., 2016; Botwe et al., 2017c). The occurrence of Sn contamination in harbour sediments is mainly attributable to the use of tributyltin (TBT)-based anti-fouling paints on ships, boats and docks (Berto et al., 2007; Nyarko et al. 2011; Castro et al. 2012; Botwe et al., 2017d). In 2001, the International Maritime Organisation (IMO) adopted the International Convention on the Control of Harmful Anti-fouling Systems on Ships, which calls on states to take steps toward the reduction of organotin pollution due to the ecological risks posed by TBT to the marine environment and human health. Since then, TBT use has been restricted in many countries (Berto et al., 2007), but could remain in use in developing countries like Ghana. Thus, the presence of Sn in the Tema Harbour sediments may derive from current use of TBT-based anti-fouling paints in the Tema Harbour or residual/secondary contaminant sources since TBT is persistent in the environment (Berto et al., 2007) and, therefore, its attenuation and recovery in sediments may be a slow process. Arsenic (As) is hazardous (Casado-Martínez et al., 2006; Affum et al., 2016; Botwe et al., 2017d) and a Class A human carcinogen (Chen, 2001) and therefore, its "heavy" contamination in the Tema Harbour sediments is of considerable concern. The increasing temporal trend of As concentrations (Fig. 8.1i) points to the existence of continuous sources of As at the harbour, which may be linked to metal smelting activities and the use of As-containing agrochemicals in Tema (El-Sorogy et al., 2016).

Significant correlations among metals in sediments have commonly been attributed to similar sources of the metals (Qu and Kelderman, 2001; Ramirez et al., 2005; Nyarko et al., 2011; Nyarko et al., 2014; Affum et al., 2016; Botwe et al., 2017d). Since the EF and Igeo values indicate Ni and Cu in the Tema Harbour sediments derived mainly from lithogenic origin, other factors such as similar accumulation patterns of the metals in sediments could also explain the observed Cu/Zn, Cu/Pb, Cu/Cr, Zn/Pb, Cr/Sn, Cr/Ni and Sn/Hg correlations in the harbour sediments (Aloupi and Angelidis, 2001). Thus, it is important to exercise caution in the use of EFs and Igeo as interpretative tools for the assessment of metal sources and contamination, especially when average crustal metal concentrations are used as proxies for background metal concentrations.

Despite potential differences in grain size distribution, geology and methodologies, we compared the Sn and As concentrations in the Tema Harbour sediments with concentrations found in other coastal sediments worldwide (Table 8.6). The As concentrations in the Tema Harbour sediments were high compared to values reported for gold mining areas in Ghana (Affum et al., 2016) and coastal sediments from Egypt (Mostafa et al., 2004; Abdel Ghani et al., 2013), Finland (Vallius et al., 2007), China (Luo et al., 2010), Spain (Díaz-de Alba et al., 2011), Tanzania (Rumisha et al., 2012), the French Mediterranean (Mamindy-Pajany et al., 2013), and Saudi Arabia (Al-Taani et al., 2014; Youssef et al., 2015; El-Sorogy et al., 2016). The concentrations of Sn in the Tema Harbour sediments were low compared to values reported for coastal sediments from Spain (Arambarri et al., 2003) and some parts of Egypt (Abdel Ghani et al., 2013), but high compared to other parts of Egypt (El-Moselhy and Hamed, 2000; Mostafa et al., 2004; Hamed et al., 2013) and China (Liu et al., 2011).

Table 8.6 Comparison of Sn and As concentrations in the Tema Harbour fine sediments investigated with concentrations in other coastal sediments worldwide (mg kg^{-1} dw)

Sediment location	Sn	As	Reference
Tema Harbour (Ghana)	6.2-49.2	146-1470	This study
Gipuzkoa Estuary (Spain)	11-113	-	Arambarri et al., 2003
Abu-Qir Bay and Eastern Harbour (Alexandria, Egypt)	3.4-355	1.6-16.2	Abdel Ghani et al., 2013
Red Sea coast (Egypt)	0.02-1.2	-	El-Moselhy and Hamed, 2000
Western Harbour (Alexandria, Egypt)	2.1-15.3	4.7-15	Mostafa et al., 2004
Pearl River Estuary (China)	1.7-8	-	Liu et al., 2011
Mediterranean coast (Egypt)	0.42-3.8	-	Hamed et al., 2013
Bonsa River (Ghana)	-	0.35-1.09	Affum et al., 2016
Gulf of Finland	-	7.3-19.1	Vallius et al., 2007
Northern Bohai and Yellow Seas (China)	-	5.6-13	Luo et al., 2010
Algeciras Bay (Spain)	-	8-23	Díaz-de Alba et al., 2011
Salaam coast (Tanzania)	-	0.2-1.3	Rumisha et al., 2012
French Mediterranean marinas	-	17-350	Mamindy-Pajany et al., 2013
Gulf of Aqaba (Saudi Arabia)	-	12.2-15.1	Al-Taani et al., 2014
Arabian Gulf (Saudi Arabia)	-	148	Youssef et al., 2015
Tarut Island (Arabian Gulf, Saudi Arabia)	-	53-342	El-Sorogy et al., 2016

8.4.2. Settling fluxes of silt-clay particles and associated metals in the Tema Harbour

The fluxes of silt-clay particles and metals in the Tema Harbour were characterised by large local fluctuations, which may be attributed to localised metal inputs and complex dynamics of sediments under the prevailing conditions in the Tema Harbour (Botwe et al., 2017b, e). In the Tema Harbour, the shear bottom stress associated with the maximum tidal currents exceeds the threshold value for resuspension of clays and fine silts (Botwe et al., 2017b) and thus, tidal and wave actions are bound to cause resuspension and redistribution of (metal-contaminated) sediments following deposition. Moreover, anthropogenic disturbances such as the movement of vessels may induce sediment resuspension with subsequent redistribution in the harbour (Lepland et al., 2010; Green and Coco, 2014). Since sediment traps integrate sediment settling and resuspension (Botwe et al., 2017b), the sediment trap-derived fluxes represent the gross downward fluxes (depositional rates) of silt-clay and associated contaminants from the water column into bottom sediments.

Although there were no clear spatial trends in the fluxes of the silt-clay particles and associated metals, relatively higher fluxes occurred at station IV. This is possibly a direct result of the high tidal current at the entrance of the Main Harbour, which has the potential to cause influxes and remobilisation of large amounts of sediment compared with the inner stations where the tidal currents are lower (Botwe et al., 2017b). Considering that station V is closer to the harbour entrance but recorded lower particle fluxes than station IV, it is also possible that the tidal currents begin to lose a significant amount of energy around station IV upon entry into the harbour. Strong direct correlations were observed among the silt-clay particles and metal fluxes indicating that metal transport to depth and its distribution in the Tema Harbour are well-regulated by settling of the silt-clay particles, which can be attributed to strong metal-silt-clay interactions in the water column (Horowitz and Elrick, 1987; Horowitz et al., 1989; Hostache et al., 2014). Thus, high settling fluxes of silt-clay in the harbour may result in substantial fluxes of metals into the bottom sediments, a situation that can have adverse ecological effects such as contaminant uptake by benthic organisms, reduced light availability, loss of spawning areas, and smothering of benthic eggs and larvae (Green and Coco, 2014).

8.4.3. Ecotoxicological risks of metal contamination in settling silt-clay particles in the Tema Harbour

Contaminated silt-clay particles represents an important pathway of contaminant exposure to the aquatic food chain as it can be easily ingested by benthic organisms (Bat and Raffaelli, 1998; Bat, 2005; Botwe et al., 2017a). A pair of empirically derived numerical sediment quality guidelines (SQGs), namely the effects-range low (ERL) and effects-range median (ERM) (Long et al., 1995; Long et al., 1998) were used to evaluate the potential ecotoxicological risks associated with metal contamination in the settling silt-clay particles in the Tema Harbour (Botwe et al., 2017a). Metal concentrations < ERL may cause rare occurrence of adverse biological effects and thus pose low potential ecotoxicological risk, concentrations > ERM may cause frequent occurrence of adverse biological effects and thus pose high potential ecotoxicological risk, whereas concentrations in the ERL-ERM range may cause infrequent occurrence of adverse biological effects and thus pose medium potential ecotoxicological risk (Long et al., 1995; Long et al., 1998). Table 8.7 shows that As concentrations in all the analysed sediments may pose high ecotoxicological risk, while the concentrations of Cr, Cu, Zn, Ni and Hg may mainly pose medium risk. On the contrary, the sediment Pb concentrations may be mainly associated with low ecotoxicological risk. SQGs for Mn and Sn are not available and therefore, their associated ecotoxicological risks cannot be evaluated using the ERL/ERM approach. Thus, although the settling fluxes of the silt-clay particles were low, they are potentially hazardous due to the high concentrations of the associated metals, particularly As and Sn, for which appropriate (bio)remediation strategies need to be developed.

Table 8.7 Comparison of metal concentrations in the settling silt-clay particles at Tema Harbour (Ghana) with effects-range low (ERL) and effects-range median (ERM) sediment quality guideline (SQG) values

SQG	SQG value (mg kg⁻¹ dw)								
	Mn	Pb	Cr	Cu	Zn	Ni	Hg	Sn	As
[a]ERL	NA	46.7	81	34	150	20.9	0.15	NA	8.2
[a]ERM	NA	218	370	270	410	51.6	0.71	NA	70
% samples below ERL	-	68	12	14	7	4	11	-	0
% samples within ERL-ERM range	-	30	86	86	93	71	77	-	0
% samples exceeding ERM	-	2	2	0	0	25	12	-	100

[a]Long et al. (1995); NA = not available

8.5. Conclusions

Settling fluxes of the silt-clay particles and their associated metals (Mn, Pb, Cr, Cu, Zn, Ni, Hg, Sn and As) in the Tema Harbour have been investigated. The mean daily fluxes ranged from 42.7 to 85.0 g m⁻² d⁻¹ for the silt-clay particles and from 15.8-49.4, 1.6-3.3, 5.7-17.6, 1.8-3.7, 6.4-14.9, 1.7-3.0, (1.3-11.4) x10⁻², 1.0-2.8, 18.5-45.2 mg m⁻² d⁻¹ for Mn, Pb, Cr, Cu, Zn, Ni, Hg, Sn and As, respectively. The silt-clay fluxes and metal fluxes were characterised by large local fluctuations possibly due to localised inputs and complex sediment dynamics in the harbour. The silt-clay fluxes and metal fluxes showed strong correlations, indicating the important role of the silt-clay particles in metal transport and distribution in the harbour. Comparison with SQGs show that the concentrations of most metals in the settling silt-clay particles are potentially toxic to benthic species and pose high ecotoxicological risks, with As posing the greatest risk. Geochemical indices revealed different extents of metal contamination in the harbour sediments, being most polluted with As and least polluted with

Mn, Cu and Ni. The As concentrations showed an increasing temporal trend, indicating the existence of continuous sources of As at the Tema Harbour, likely linked to metal smelting and the use of As-containing agrochemicals in the area. This study shows that the Tema Harbour is increasingly being impacted by anthropogenic activities. The concentrations of As and Sn are of concern, which calls for increased efforts to address the contamination problem.

References

Abdel Ghani, S., El Zokm, G., Shobier, A., Othman, T., & Shreadah, M. (2013). Metal pollution in surface sediments of Abu-Qir Bay and Eastern Harbour of Alexandria, Egypt. *The Egyptian Journal of Aquatic Research, 39*(1), 1-12.

Addo, M.A., Okley, G., Affum, H., Acquah, S., Gbadago, J., Senu, J., & Botwe, B.O. (2011). Water quality and level of some heavy metals in water and sediments of Kpeshie Lagoon, La-Accra, Ghana. *Research Journal of Environmental and Earth Sciences, 3*(5), 487-497.

Affum, A.O., Osae Dede, S., Nyarko, B.J.B, Acquaah, S., Kwaansa-Ansah, E.E, Darko, G., Dickson, A., Affum, E., & Fianko, J.R. (2016). Influence of small-scale gold mining and toxic element concentrations in Bonsa river, Ghana: A potential risk to water quality and public health. *Environmental Earth Sciences, 75*, 178-194.

Al-Taani, A.A., Batayneh, A., Nazzal, Y., Ghrefat, H., Elawadi, E., & Zaman, H. (2014). Status of trace metals in surface seawater of the Gulf of Aqaba, Saudi Arabia. *Marine Pollution Bulletin, 86*(1), 582-590.

Arambarri, I., Garcia, R., & Millán, E. (2003). Assessment of tin and butyltin species in estuarine superficial sediments from Gipuzkoa, Spain. *Chemosphere, 51*(8), 643-649.

Bat, L. (2005). A review of sediment toxicity bioassays using the amphipods and polychaetes. *Turkish Journal of Fisheries and Aquatic Sciences, 5*(2), 119-139.

Bat, L., & Raffaelli, D. (1998). Sediment toxicity testing: a bioassay approach using the amphipod *Corophium volutator* and the polychaete *Arenicola marina*. *Journal of Experimental Marine Biology and Ecology, 226*(2), 217-239.

Berto, D., Giani, M., Boscolo, R., Covelli, S., Giovanardi, O., Massironi, M., & Grassia, L. (2007). Organotins (TBT and DBT) in water, sediments, and gastropods of the southern Venice lagoon (Italy). *Marine Pollution Bulletin, 55*(10), 425-435.

Bloesch, J., & Burns, N. (1980). A critical review of sedimentation trap technique. *Schweizerische Zeitschrift für Hydrologie, 42*(1), 15-55.

Botwe, B.O., De Schamphelaere, K., Schipper, C.A., Teuchies, J., Blust, R., Nyarko, E., & Lens, P.N.L. (2017a). Integrated hazard, risk and impact assessment of tropical marine sediments from Tema Harbour (Ghana). *Chemosphere, 177*, 24-34.

Botwe, B.O., Abril, J.M., Schirone, A., Barsanti, M., Delbono, I., Delfanti, R., Nyarko, E., & Lens, P.N.L. (2017b). Settling fluxes and sediment accumulation rates by the combined use of sediment traps and sediment cores in Tema Harbour (Ghana). *Science of the Total Environment, 609*, 1114-1125.

Botwe, B.O., Kelderman, P., Nyarko, E., & Lens, P.N.L. (2017c). Assessment of DDT, HCH and PAH contamination and associated ecotoxicological risks in surface sediments of coastal Tema Harbour (Ghana). *Marine Pollution Bulletin, 115*(1–2), 480-488.

Botwe, B.O., Alfonso, L., Nyarko, E., & Lens, P.N.L. (2017d). Metal distribution and fractionation in surface sediments of coastal Tema Harbour (Ghana) and its ecological implications. *Environmental Earth Sciences, 76*(15), 514-530

Botwe, B.O., Schirone, A., Delbono, I., Barsanti, M., Delfanti, R., Kelderman, P., Nyarko, E., & Lens, P.N.L. (2017e). Radioactivity concentrations and their radiological significance in sediments of the Tema Harbour (Greater Accra, Ghana). *Journal of Radiation Research and Applied Sciences, 10*(1), 63-71.

Buesseler, K.O., Antia, A.N., Chen, M., Fowler, S.W., Gardner, W.D., Gustafsson, O., Harada, K., Michaels, A.F., Rutgers van der Loeff, M., & Sarin, M. (2007). An assessment of the use of sediment traps for estimating upper ocean particle fluxes. *Journal of Marine Research, 65*(3), 345-416.

Canadian National Laboratory for Environmental Testing (1994). Manual of analytical methods. Trace Metals, 2:1-14

Casado-Martínez, M.C., Buceta, J.L., Belzunce, M.J., & DelValls, T.A. (2006). Using sediment quality guidelines for dredged material management in commercial ports from Spain. *Environment International, 32*(3), 388-396.

Castro, Í.B., Arroyo, M.F., Costa, P.G., & Fillmann, G. (2012). Butyltin compounds and imposex levels in Ecuador. *Archives of Environmental Contamination and Toxicology, 62*(1), 68-77.

Chen, M., Ma, L.Q., Hoogeweg, C.G., & Harris, W.G. (2001). Arsenic background concentrations in Florida, U.S.A. surface soils: Determination and interpretation. *Environmental Forensics, 2*(2), 117-126.

Clark, R.B. (2001). Marine Pollution. Oxford: Oxford University Press.

de Vicente, I., Cruz-Pizarro, L., & Rueda, F.J. (2010). Sediment resuspension in two adjacent shallow coastal lakes: controlling factors and consequences on phosphate dynamics. *Aquatic Sciences, 72*(1), 21-31.

Díaz-de Alba, M., Galindo-Riaño, M.D., Casanueva-Marenco, M.J., García-Vargas, M., & Kosore, C.M. (2011). Assessment of the metal pollution, potential toxicity and speciation of sediment from Algeciras Bay (South of Spain) using chemometric tools. *Journal of Hazardous Materials, 190*(1), 177-187.

Eisma, D. (1993). Suspended Matter in the Aquatic Environment. Springer-Verlag.

El-Moselhy, K., & A. Hamed, M. (2000). Pollution by mercury and tin in water and sediments of the Red Sea coasts, Egypt. *Journal of Environmental Sciences, 19*, 29-41.

El-Sorogy, A.S., Youssef, M., Al-Kahtany, K., & Al-Otaiby, N. (2016). Assessment of arsenic in coastal sediments, seawaters and molluscs in the Tarut Island, Arabian Gulf, Saudi Arabia. *Journal of African Earth Sciences, 113*, 65-72.

Green, M.O., & Coco, G. (2014). Review of wave-driven sediment resuspension and transport in estuaries. *Reviews of Geophysics, 52*(1), 77-117.

Hakanson, L. (2006). Suspended Particulate Matter in lakes, rivers, and marine systems. Blackburn Press, Caldwell.

Hakanson, L., & Jansson, M. (1983) Principles of lake sedimentology. Springer Verlag, Berlin.

Hamed, M.A., Mohamedein, L. I., El-Sawy, M.A., & El-Moselhy, K.M. (2013). Mercury and tin contents in water and sediments along the Mediterranean shoreline of Egypt. *The Egyptian Journal of Aquatic Research, 39*(2), 75-81.

Hamzeh, M., Ouddane, B., Daye, M., & Halwani, J. (2014). Trace metal mobilization from surficial sediments of the Seine River estuary. *Water, Air, & Soil Pollution, 225*(3), 1878-1892.

Helali, M.A., Zaaboub, N., Oueslati, W., Added, A., & Aleya, L. (2016). Suspended particulate matter fluxes along with their associated metals, organic matter and carbonates in a coastal Mediterranean area affected by mining activities. *Marine Pollution Bulletin, 104*(1), 171-181.

Horowitz, A.J. (1985). A primer on trace metal-sediment chemistry, U.S. Geological Survey Water-Supply Paper 2277, 67 p.

Horowitz, A.J., & Elrick, K.A. (1987). The relation of stream sediment surface area, grain size and composition to trace element chemistry. *Applied Geochemistry, 2*(4), 437-451.

Horowitz, A.J., Elrick, K.A., & Hooper, R.P. (1989). The prediction of aquatic sediment-associated trace element concentrations using selected geochemical factors. *Hydrological Processes, 3*(4), 347-364.

Hostache, R., Hissler, C., Matgen, P., Guignard, C., & Bates, P. (2014). Modelling suspended-sediment propagation and related heavy metal contamination in floodplains: a parameter sensitivity analysis. *Hydrology & Earth System Sciences, 18*(9), 3539-3551.

Ji, Z.G. (2008). Hydrodynamics and Water Quality: Modelling Rivers, Lakes and Estuaries. Wiley.

Kelderman, P., Ang'weya, R., De Rozari, P., & Vijverberg, T. (2012). Sediment characteristics and wind-induced sediment dynamics in shallow Lake Markermeer, the Netherlands. *Aquatic Sciences, 74*(2), 301-313.

Lepland, A., Andersen, T.J., Lepland, A., Arp, H.P.H., Alve, E., Breedveld, G.D., & Rindby, A. (2010). Sedimentation and chronology of heavy metal pollution in Oslo harbor, Norway. *Marine Pollution Bulletin, 60*(9), 1512-1522.

Liu, B., Hu, K., Jiang, Z., Yang, J., Luo, X., & Liu, A. (2011). Distribution and enrichment of heavy metals in a sediment core from the Pearl River Estuary. *Environmental Earth Eciences, 62*(2), 265-275.

Liu, J., Clift, P.D., Yan, W., Chen, Z., Chen, H., Xiang, R., & Wang, D. (2014). Modern transport and deposition of settling particles in the northern South China Sea: Sediment trap evidence adjacent to Xisha Trough. *Deep Sea Research Part I: Oceanographic Research Papers, 93*, 145-155.

Long, E.R., Field, L.J., & MacDonald, D.D. (1998). Predicting toxicity in marine sediments with numerical sediment quality guidelines. *Environmental Toxicology and Chemistry, 17*(4), 714-727.

Long, E.R., MacDonald, D.D., Smith, S. L., & Calder, F.D. (1995). Incidence of adverse biological effects within ranges of chemical concentrations in marine and estuarine sediments. *Environmental Management, 19*(1), 81-97.

Luo, W., Lu, Y., Wang, T., Hu, W., Jiao, W., Naile, J.E., Khim, J.S., & Giesy, J.P. (2010). Ecological risk assessment of arsenic and metals in sediments of coastal areas of northern Bohai and Yellow Seas, China. *Ambio, 39*(5-6), 367-375.

Mamindy-Pajany, Y., Bataillard, P., Séby, F., Crouzet, C., Moulin, A., Guezennec, A.-G., Hrel, C., Marmier, N., & Battaglia-Brunet, F. (2013). Arsenic in Marina Sediments from the Mediterranean Coast: Speciation in the Solid Phase and Occurrence of

Thioarsenates. *Soil and Sediment Contamination: An International Journal, 22*(8), 984-1002.

Mestres, M., Sierra, J., Mösso, C., & Sánchez-Arcilla, A. (2010). Sources of contamination and modelled pollutant trajectories in a Mediterranean harbour (Tarragona, Spain). *Marine Pollution Bulletin, 60*(6), 898-907.

Mostafa, A.R., Barakat, A.O., Qian, Y., Wade, T.L., & Yuan, D. (2004). An overview of metal pollution in the Western Harbour of Alexandria, Egypt. *Soil and Sediment Contamination: An International Journal, 13*(3), 299-311.

Muller, G., 1969. Index of geo-accumulation in sediments of the Rhine River. *Journal of Geology, 2*(3): 108–118.

Nyarko, E., Botwe, B.O., Lamptey, E., Nuotuo, O., Foli, B.A., & Addo, M.A. (2011). Toxic metal concentrations in deep-sea sediments from the jubilee oil field and surrounding areas off the western coast of Ghana. *Tropical Environmental Research, 9*, 584-595.

Nyarko, E., Fletcher, A., Addo, S., Foli, B.A.K., & Mahu, E. (2014). Geochemical assessment of heavy metals in surface sediments: a case study of the Tema Port, Ghana. *Journal of Shipping & Ocean Engineering, 4, 79-92.*

Petrosillo, I., Valente, D., Zaccarelli, N., & Zurlini, G. (2009). Managing tourist harbors: Are managers aware of the real environmental risks? *Marine Pollution Bulletin, 58*(10), 1454-1461.

Pozza, M.R., Boyce, J.I., & Morris, W.A. (2004). Lake-based magnetic mapping of contaminated sediment distribution, Hamilton Harbour, Lake Ontario, Canada. *Journal of Applied Geophysics, 57*(1), 23-41.

Qu, W., & Kelderman, P. (2001). Heavy metal contents in the Delft canal sediments and suspended solids of the River Rhine: multivariate analysis for source tracing. *Chemosphere, 45*(6–7), 919-925.

Ramirez, M., Massolo, S., Frache, R., & Correa, J.A. (2005). Metal speciation and environmental impact on sandy beaches due to El Salvador copper mine, Chile. *Marine Pollution Bulletin, 50*(1), 62-72.

Rumisha, C., Elskens, M., Leermakers, M., & Kochzius, M. (2012). Trace metal pollution and its influence on the community structure of soft bottom molluscs in intertidal areas of the Dar es Salaam coast, Tanzania. *Marine Pollution Bulletin, 64*(3), 521-531.

Santos-Echeandía, J., Prego, R., Cobelo-García, A., & Caetano, M. (2012). Metal composition and fluxes of sinking particles and post-depositional transformation in a ria coastal system (NW Iberian Peninsula). *Marine Chemistry, 134*, 36-46.

Schipper, C., Rietjens, I., Burgess, R., & Murk, A. (2010). Application of bioassays in toxicological hazard, risk and impact assessments of dredged sediments. *Marine Pollution Bulletin, 60*(11), 2026-2042.

Szmytkiewicz, A., & Zalewska, T. (2014). Sediment deposition and accumulation rates determined by sediment trap and ^{210}Pb isotope methods in the Outer Puck Bay (Baltic Sea). *Oceanologia, 56*(1), 85-106.

Taylor, S. (1964). Abundance of chemical elements in the continental crust: a new table. *Geochimica et Cosmochimica Acta, 28*(8), 1273-1285.

Vallius, H., Ryabchuk, D., Kotilainen, A. (2007). Distribution of heavy metals and arsenic in soft surface sediments of the coastal area off Kotka, northeastern Gulf of Finland, Baltic Sea. In: Vallius, H. (Ed.), Holocene Sedimentary Environment and Sediment Geochemistry of the Eastern Gulf of Finland, Baltic Sea, Geological Survey, Finland, 45:33-48.

Wang, X.-C., Zhang, Y.-X., & Chen, R.F. (2001). Distribution and partitioning of polycyclic aromatic hydrocarbons (PAHs) in different size fractions in sediments from Boston Harbor, United States. *Marine Pollution Bulletin, 42*(11), 1139-1149.

Yilgor, S., Kucuksezgin, F., & Ozel, E. (2012). Assessment of metal concentrations in sediments from Lake Bafa (Western Anatolia): an index analysis approach. *Bulletin of Environmental Contamination and Toxicology, 89*(3), 512-518.

Youssef, M., El-Sorogy, A., Al Kahtany, K., & Al Otiaby, N. (2015). Environmental assessment of coastal surface sediments at Tarut Island, Arabian Gulf (Saudi Arabia). *Marine Pollution Bulletin, 96*(1), 424-433.

Chapter 9

General discussion and conclusions

9.1. Sediment pollution and accumulation in habours

Sediment pollution is a global issue that threatens environmental sustainability and human health. Coastal marine environments, particularly harbours, are potential hot-spots of sediment pollution, as they are bound to receive considerable amounts of chemical contaminants from intense maritime and other anthropogenic activities such as industrialisation, agriculture and urbanisation (Smith et al., 2009; Petrosillo et al., 2009; Lepland et al., 2010; Mestres et al., 2010; Schipper et al., 2010; Romero et al., 2014). Moreover, water exchanges within coastal harbours can result in the influx of large amounts of suspended sediments, which settle and accumulate within harbour basins under low hydrodynamic conditions (Lepland et al., 2010; Luo et al., 2010; Mestres et al., 2010). These sediments may derive from varied sources, but are mainly supplied by riverine discharges. Globally, the discharge of sediments by rivers is estimated at 2.3 billion tonnes per year (Syvitski et al., 2005). In Ghana, rivers discharge about 2.4 million tonnes of sediment per year into the sea (Akrasi, 2011). Since sediments play an important role in the fate of chemical pollutants in aquatic systems (Mulligan et al., 2001; Ruiz-Fernandez et al 2009; Prato et al., 2011; Jiang et al., 2013), the accumulation of sediments in harbour basins can result in the accumulation of chemical pollutants.

Considering the important role harbours play in the socio-economic development of human societies, sediment pollution, along with high sediment accumulation rates (SARs) in harbours imposes remedial measures, often dredging, to ensure harbour sustainability. Annually, several millions of tonnes of sediments are dredged from harbours around the world at huge cost and subsequently disposed of at sea (Witt et al., 2004; Bolam et al., 2006; Schipper et al., 2010). Alternatively, dredged materials may be disposed of on land for agriculture, landfill, beach nourishment and restoration of wetlands (Choueri et al., 2009). Apart from the high cost of dredging, the potential of dredged materials to cause detrimental effects on receiving ecosystems has been recognised (Hong et al., 1995; Burton, 2002; Caille et al., 2003; Stronkhorst and van Hattum, 2003; Casado-Martinez et al., 2006; Birch and Hutson, 2009; Choueri et al 2009; Schipper et al., 2010).

Global concerns about the potential adverse impact of dredged sediments from harbours and their subsequent disposal on the marine environment has led to the establishment of international treaties such as the London Convention (1972) and the London Protocol (1996), which impose rigorous assessment of dredged materials in order to minimise their adverse impacts. Thus, chemical pollution and sediment accumulation rates in harbours are of wide interest to harbour engineers and managers. Many developed countries including The Netherlands, United States, Canada, Australia, New Zealand and Spain have therefore established regulatory standards for dredged materials based on potentially hazardous concentrations of chemical pollutants to guide the management of dredged material (Burton, 2002; Casado-Martinez et al., 2006; Birch and Hutson, 2009; Kelderman et al., 2012). These standards may be referred to as Sediment Quality Guidelines (SQGs) or Action Levels (ALs) in different countries of origin, and they are used in ecotoxicological risk assessment for evaluating the likelihood of a sediment contaminant to elicit adverse biological effects.

In Ghana, no national regulatory standards have been developed to guide the management of dredged materials. In this thesis, a pair of SQGs namely the effects-range low (ERL) and the effects-range median (ERM) (Long et al., 1995) were used to predict the likelihood of ecotoxicological effect due to the measured sediment contaminants. Although the ERL and ERM have been found to have good predictive ability (Long et al., 1995; Long et al., 1998), it is worthy to note that SQGs have several limitations, and have been found to result in false positive and false negative predictions in some instances (Burton, 2002). Moreover, SQGs are chemical specific and do not establish causality where mixtures of chemicals are present (Burton, 2002). Nonetheless, they are useful for identifying contaminants of potential concern (Burton, 2002; Long et al., 2006; Birch and Hutson 2009) and should be used in a "screening-level" manner or "weight-of-evidence" approach (Burton, 2002).

9.1.1. Chemical pollution assessment of Tema Harbour sediments

In this thesis, a "weight-of-evidence" approach was adopted to pollution assessment of the Tema Harbour sediments. A screening-level assessment was first conducted to provide a first line of evidence of chemical pollution in the harbour sediments, characterise the potential sources of the pollutants and identify those of potential concern (Burton, 2002; Long et al., 2006). Based on the results from the screening-level assessment, bioassays were conducted as a further line of evidence of chemical pollution of the Tema Harbour sediments.

9.1.1.1. Screening-level assessment of chemical pollution in the Tema Harbour sediments

Radiological hazard indices (Chapter 3), SQGs (Chapters 4 and 5), total toxicity equivalence (Chapter 4) and risk assessment code (Chapter 5) were used as screening tools for the assessment of chemical pollution in the Tema Harbour sediments. This screening-level assessment classified the Tema Harbour sediments as potentially toxic due to DDTs, HCHs and PAHs (Chapter 4), As, Cd, Hg and Zn (Chapter 5) contamination. γ-HCH and Hg were identified as contaminants of potential concern in Chapters 4 and 5, respectively. This first line of evidence necessitated a more refined assessment (Burton, 2002; Long et al., 2006), which will be discussed shortly.

To assess anthropogenic influences on harbours, many researchers have analysed the deep and surficial layers of sediment cores (Birch and Olmos, 2008; Abrahim and Parker, 2008; Yilgor et al., 2012). The deep layers of sediment cores represent materials deposited before the construction of the harbour and may thus reveal the pre-impacted environmental conditions for the harbour area. The surficial layers of sediment cores, on the other hand, represent those that were deposited after the construction of the harbour and occur closer to the sediment-water interface. Therefore, contaminant levels in the deep sediment layers of sediment cores can be used to establish background levels for the harbour area, against which the levels in the surficial sediments can be compared to assess the extent of contamination. This approach is ideal considering that background levels of chemical contaminants may vary regionally due to potential differences in geological properties (Jiang et al., 2013).

Chapter 7 showed that the lengths of sediment cores collected from the Tema Harbour were too short and only captured the past 13-24 years of sedimentation. Considering that the harbour has been in existence for over fifty years, the sediment cores could not be used to establish contaminant background levels in the Tema Harbour. Moreover, no consistent sediment contamination monitoring data have been collected for the harbour over the years. Therefore, for the assessment of metal contamination levels and sources in the Tema Harbour (Chapters 5 and 8), two widely used quantitative geochemical indices, namely the metal enrichment factor (EF) proposed by Sinex and Helz (1981) and the geo-accumulation index (I_{geo}) proposed by Muller (1969) were applied. These geochemical indices were used to characterise the extent of metal enrichment/contamination and apportion their sources as natural or anthropogenic, adopting the average crustal concentrations of metals reported by Taylor (1964) as proxies for the metal background levels. Although this has become a standard approach used by many authors (e.g. Addo et al., 2012; Mahu et al., 2015), there is a need to exercise caution in the interpretation of the results.

For the characterisation of the dominant sources of PAH contamination in sediments, the ΣLPAH/ΣHPAH ratio (Rocher et al., 2004) and various PAH isomeric ratios including anthracene/[anthracene + phenanthrene], fluoranthene/[fluoranthene + pyrene] and benzo(a)anthracene/[benzo(a)anthracene + chrysene] have been widely used (Yunker et al., 2002; Abrajano et al., 2003; Rocher et al., 2004; Nyarko et al., 2011; Guerra, 2012). Chapter 4 showed that these PAH source diagnostic tools did not yield consistent results for the Tema Harbour sediments. This is an interesting finding, which suggests that a number of indices should be used when assessing PAH sources in sediments.

9.1.1.2. *Bioassay testing of Tema Harbour sediment*

Results from Chapters 4 and 5 indicated potential toxicity of the harbour sediments, which required a more refined assessment. Whole-sediment toxicity bioassays are recommended as a valuable tool in ecotoxicological assessment of contaminated sediments (Annicchiarico et al., 2007; Ré et al., 2009; Schipper et al., 2010) as they integrate the toxic effects of contaminant mixtures present in a sediment sample (Forrester et al., 2003; Escher et al., 2008). In many countries, bioassays are now required in the assessment of dredged materials and for the licensing of their subsequent disposal (DelValls et al., 2004; Rose et al., 2006; Casado-Martinez et al., 2006, 2007; van Dam et al., 2008; Prato et al., 2011). Thus, in Chapter 6, the standard acute 10-day *C. volutator* (Roddie and Thain, 2002; Schipper et al., 2006) and the standard chronic 28-day *H. diversicolor* (Hannewijk et al., 2004) whole-sediment bioassays were combined with chemical analysis, SQGs and ALs to characterise the potential hazard/toxicity of the Tema Harbour sediments as a further line of evidence of sediment pollution in the Tema Harbour. For the bioassay, mortality and metal bioaccumulation were evaluated as endpoints.

The *C. volutator* and *H. diversicolor* whole-sediment bioassay tests were chosen for the following reasons (Ciarelli et al., 1998; Uwadiae, 2010; Carvalho et al., 2012). First, these protocols are standard (Roddie and Thain, 2002; Schipper et al., 2006; Hannewijk et al., 2004). Secondly, they have been widely used in the scientific literature (Stronkhorst et al., 2003; Scarlett et al., 2007; van den Heuvel-Greve et al., 2007; Moreira et al., 2006; Mayor et al., 2008), but rarely as complementary test species. Thirdly, they are abundant, available throughout the year and, easily accessible and maintained under laboratory conditions (Scaps, 2002; Bat, 2005). Fourthly, they are known to tolerate a wide range of salinities, temperatures, sediment grain sizes and organic carbon contents (Ciarelli et al., 1998; Roddie and Thain, 2002; Scaps, 2002; Bat, 2005; Philippe et al., 2008). Lastly, they have wide geographic distributions across polar, temperate and tropical marine regions (Bat, 2005; Moreira et al., 2006; Uwadiae, 2010; Carvalho et al., 2012). In the case of the *C. volutator* whole-sediment toxicity bioassay, a sediment toxicity classification scheme has been developed (EPA, 1998).

Although the use of local species from the study area as bioassay test organism is prefereable, a biological baseline survey conducted on benthic species in the Tema Harbour in 2010 by Nyarko et al. (unpublished) revealed the absence of *C. valuator* and *H. diversicolor* in the harbour. Moreover, whole-sediment toxicity bioassays with tropical species are not yet well developed (Adams and Stauber, 2008). Therefore, for this thesis, whole-sediment toxicity bioassay tests on the tropical Tema Harbour sediments were conducted in Europe, using *C. volutator* and *H. diversicolor* of European origin (Chapter 6).

The bioassay tests showed that the Tema Harbour sediments are hazardous as predicted by the SQGs and the TEQs (Table 4.5; Table 5.4). Thus, the absence of these species at the Tema Harbour may partly be due to their vulnerability to the toxicity of the harbour sediments. Chapters 3-5 indicated that the bottom water within the harbour was anoxic, which could also be unfavourable for *C. valuator* and *H. diversicolor* to thrive. The pH range (7.3-8.3) of the harbour sediments (Chapter 5) was, however, comparable to the pH range (7.7-8.5) of the overlying water in the bioassays (Chapter 6). Different benthic species may display different sensitivities to sediment toxicity (Dauvin et al., 2016) and this was observed, *C. volutator* being more sensitive than *H. diversicolor* (Figs. 6.2 and 6.3). The difference in the sensitivities may be partly due to the presence of different toxicants in the harbour sediments (Bat and Raffaelli, 1998; Forrester et al., 2003; Long et al., 2006; Mayor et al., 2008), i.e. radionuclides (Chapter 3), DDTs, HCHs and PAHs (Chapter 4) and metals (Chapters 5 and 6). Based on SQGs, the sediment concentrations of Pb, Ni, Cu, Zn and As may pose medium potential ecotoxicological risks (Table 6.5). A logarithmic relationship was observed between *C. volutator* mortality and sediment Cd concentration (Fig. 6.4a), whereas a linear relationship was observed between *H. diversicolor* mortality and sediment Cu concentration (Fig. 6.4b), suggesting Cd and Cu may play a role in the mortality of *C. valuator* and *H. diversicolor*, respectively.

Amphipods are generally highly sensitive to low levels of hydrocarbon contamination, whereas polychaetes are more tolerant (Dauvin et al., 2016). During the bioassay experiments, petroleum contamination was visible in the sediments from the Fishing Harbour and the Canoe Basin, possibly as a result of oil spills during refuelling of fishing vessels and fuel leakage in these sampling areas (Chapter 6). Thus, PAH contamination could potentially contribute to the toxicity in the Tema Harbour sediments. This requires further investigation using appropriate sediment sampling techniques for PAH assessment. For this bioassay study, the sediments were sampled in accordance with metal analysis.

Sediment toxicity to *C. volutator* and *H. diversicolor* may occur as a result of contaminant uptake and subsequent bioaccumulation (Luoma and Rainbow, 2008; Rainbow et al., 2009). The contaminated harbour sediments were predominantly composed of the silt-clay fraction, which can be easily ingested by the test organisms (Bat and Raffaelli, 1998; Burton, 2002). Therefore, to provide a further line of evidence of the potential biological effects of sediment pollution in the Tema Harbour, Chapter 6 also evaluated the bioaccumulation potential of sediment-bound metals (Cd, Pb, Cr, Ni, Cu, Zn and As) in the exposed *C. volutator* and *H. diversicolor* based on the biota-sediment bioaccumulation factors (BSAFs) (Aydin-Onen et al., 2015).

Compared with the control organisms, it was evident that metal bioaccumulation occurred in both *C. volutator* and *H. diversicolor* exposed to the Tema Harbour sediments. This indicates high potential bioavailability and risk of metals entering the food. Comparison of metal concentrations in the Tema Harbour sediments with international Action Levels (Table 6.6) indicated unacceptable levels of Cu and Zn in the harbour sediments. This, together with the bioassay results, is of concern considering the adverse implications for the aquatic food web and the disposal of dredged material at sea without any prior remediation at the Tema Harbour. In year 2013, the Ghana Ports and Harbours Authority embarked on a port expansion project aimed at increasing the vessel handling capacity at the Tema Harbour to about 1,000,000 Twenty-Foot Equivalent Units (TEUs) in order to meet the growing needs of the international maritime market (http://ghanaports.gov.gh/GPHA). Since then, undisclosed quantities of sediments have been dredged and disposed of at sea (P.K. Ofori-Danson, pers. comm.).

In this study, the metal concentrations in the tissues of both *C. volutator* and *H. diversicolor* neither correlated statistically with the sediment total metal concentrations nor the exchangeable metal fraction in the sediments, which poses the greatest risk of entering the food chain (Jain, 2004; van Hullebusch et al., 2005; Dung et al., 2013; Kalwa et al., 2013; Pini et al., 2015). These non-correlations suggest mediation of metal bioaccumulation by processes such as detoxification (Rainbow and Luoma, 2011), storage or elimination in the organisms (Adams et al., 2011).

9.1.2. Assessment of SARs in harbours

Sediment accumulation rates in harbours are a key factor that determines the timing of dredging and the quantity of material to dredge. Information on sedimentation processes and accurate quantification of SARs in harbours is therefore essential for harbour management (Buesseler et al., 2007; Leys and Mulligan, 2011). The development of the ^{210}Pb sediment dating technique has contributed to our understanding of the dynamics and functioning of aquatic systems, with the unique ability to provide insight into the natural variability in environmental conditions as well as anthropogenic impacts in sedimentary systems on time scales spanning 100-150 years (Lu, 2007; Appleby, 2008; Díaz-Asencio et al., 2009). The ^{210}Pb dating technique can be applied in establishing sediment chronology in lacustrine, estuarine and marine environments, based on which sedimentation rates can be evaluated and pollution history reconstructed.

Deriving accurate sediment chronology and evaluating sediment accumulation rates from ^{210}Pb profiles in sediment cores is, however, not a simple and straight-forward task as the usual assumptions involved in this technique could not be accomplished in perturbed environments. The delivery of sediments and ^{210}Pb to aquatic systems is controlled by complex processes which may not be well understood. Moreover, sediment mixing is a major constraint in the quatification of recent SARs in disturbed environments such as harbours since undisturbed sediments are the optimal materials for the applicability of the widely used ^{210}Pb sediment dating models (Santschi et al., 2001; Bellucci et al., 2012).

In harbours, the movement of vessels and other activities such as dredging often induce sediment resuspension and mixing (Lepland et al., 2010; Leys and Mulligan, 2011; Green and Coco, 2014), disrupting natural sedimentation processes such as deposition (Klubi et al.,

2017) and the original imprint in the sediments (Santschi et al., 2001). Therefore, when assessing sedimentation in harbours, the issue of mixing is quite pertinent. Mixing of particle-bound tracers implies that the grain particles within the sediments are reworked, and the causal mechanisms (physical, chemical and/or biological) must be identified, as well as their depth-dependence. Thus, some authers incorporate sediment imaging techniques such as X-radiography (Schmidt et al., 2007; Bellucci et al 2010, 2012; Lepland et al 2010) and magnetic measurements (Bellucci et al., 2012) into the sediment core dating process to assess sediment mixing. X-radiographic imaging and magnetic measurements are useful tools for elucidating the sediment stratigraphy and variability in sediment texture and chemistry (Schmidt et al., 2007a).

9.1.2.1. *Assessment of SARs in the Tema Harbour*

In this thesis (Chapter 7), a rigorous approach involving sediment traps, sediment cores and the use of multitracers (^{7}Be, ^{234}Th and ^{210}Pb) and different radiometric models was employed to investigate suspended particulate matter (SPM) dynamics and quantify recent SARs in the Tema Harbour. All the sediment cores were sampled out of the dredged area in the Tema Harbour and were expected to have preserved a sequence of continuous sediment deposition. Sediment mixing was examined by visual inspection of the core surfaces for biological activity and analyses of short-lived radionuclides (^{7}Be and ^{234}Th) profiles in the upper layers of the sediment cores (Erten, 1997; Schmidt et al., 2007a, b). Due to their short half-lives, ^{234}Th and ^{7}Be can provide reliable information on sediment mixing or particle reworking processes over a period of 4-8 months (Schmidt et al., 2007a). Subsurface occurrence of ^{234}Th and ^{7}Be in areas of low sedimentation rates indicates post-depositional sediment mixing (Schmidt et al., 2007b). Visual examination of the cores revealed no evidence of bioturbation in the surficial layers, which can cause sediment mixing over large sediment mass depths. In addition, specific activities of ^{7}Be and ^{234}Th were measured in the top layers of the sediment cores to serve as a quality test for the complete surface recovery of the core, which is necessary for dating (Erten et al., 1985; Erten, 1997). The depth profiles of these radionuclides showed sharp gradients in the upper 2 cm layers (Table 7.2). Since this observation is not consistent with a well-mixed layer, the hypothesis of mixing was rejected (Section 7.3.4).

The sediment cores exhibited variable bulk density profiles (Fig. 7.2), from which highly dynamic and non-steady sedimentation conditions were inferred. The sediment trap-derived settling fluxes revealed the existence of local disturbances with a highly irregular spatial and temporal character, which pointed to a complex dynamics of rising clouds of SPM and settling in the Tema Harbour (Section 7.3.1).

9.1.2.1.1. *Quantification of recent SARs in the Tema Harbour based on ^{7}Be and ^{234}Th profiles and conventional ^{210}Pb dating models*

^{7}Be, ^{234}Th and ^{210}Pb are useful complementary tracers for studying sedimentation processes (Smoak and Patchineelam, 1999; Smoak et al., 1999; Yeager et al., 2005). Studies have shown that it is necessary to recover sedimentation records of at least five decades in order to calculate ^{210}Pb-derived sedimentation rates and nuclide inventories more rigorously (Su and Huh, 2002). The analysed cores were too short (up to $\approx$50 cm) to allow any reliable estimation of the total ^{210}Pb$_{exc}$ inventories required for the application of the constant rate of supply (CRS) model (Fig. 7.5). Moreover, a clear monotonic exponential trend of decrease was absent in the cores and thus, the application of the CF-CS model for deriving SAR values was not reliable. The low values and large uncertainties in the ^{137}Cs data did not allow any proper identification of chronostratigraphic horizons in the sediment cores, hence restricting its use as a tracer for sedimentation (Pfitzner et al., 2004).

Under some simplifying assumptions of the Constant Flux Constant Sedimentation (CF-CS) model, the ^{7}Be profiles were used to obtain a first estimate of the order of magnitude of very recent SARs over the past 6-8 months in the Tema Harbour ranging from 2.5 to 9.0 ($\pm$ 1.3) g.cm^{-2}.y^{-1} (Table 7.3). The ^{7}Be-derived SAR values, along with the ^{210}Pb$_{exc}$ specific activities in the upper sediment layers, was then used to obtain a first estimate of the ^{210}Pb$_{exc}$ fluxes onto the SWI for one core (core A = 25.6 kBq.m^{-2}.y^{-1}) (Section 7.3.4). This value is two orders of magnitude higher than the expected atmospheric deposition of ^{210}Pb$_{exc}$ in the Tema Harbour area (typically in the range of 100-200 Bq.m^{-2}.y^{-1}) (Section 7.3.4). The SAR values estimated from the ^{7}Be data were consistent with the ^{234}Th activity versus mass depth profiles, which showed a near-constant background level along the cores at the 2-3 cm upper sediment slices (Table 7.2).

9.1.2.1.2. Estimation of SARs from the ^{210}Pb-based TERESA model

Despite the potential of the TERESA model (Abril, 2016), literature survey showed that it has not yet been explored for dating sediment cores from disturbed sedimentary environments. The model hypothesis of varying, but statistically correlated, fluxes of matter and ^{210}Pb$_{exc}$ onto the SWI seemed reasonable for the Tema Harbour and thus, Chapter 7 explored the applicability of the TERESA model (Abril, 2016) for deriving recent SARs in the harbour. The TERESA model produced a sharp discontinuity in radionuclides profiles at the mass-depth of 4.06 g.cm^{-2} (Section 7.3.6), which confirmed the hypothesis of negligible post depositional processes. The TERESA model also produced a reasonable fit to the data and generated SARs of 1.4-3.0 g.cm^{-2}.y^{-1} and accretion rates of 1.7-3 cm.y^{-1} for the Tema Harbour (Section 7.3.9).

Although the TERESA model has been validated against synthetic cores and real data from varved sediments, this is the first time it has been applied to date sediment cores from a harbour. Other investigators have used integrated approaches to evaluate sedimentation rates in harbours (Tang et al., 2008; Smith et al., 2009; Lepland et al., 2010; Yussof et al., 2015) and disturbed bays (Bellucci et al., 2012), which could not be possible with the sole application of ^{210}Pb models. The time-averaged SAR values for the Tema Harbour by the TERESA model (1.4-3.0 g.cm^{-2}.y^{-1}) were very high in comparison to most of the data in the literature for lacustrine and coastal environments from other climatic regions. The accretion rates (1.7-3 cm.y^{-1}) were higher when compared with 0.57 cm.y^{-1} reported by Yussof et al. (2015) for the Tanjung Pelepas Harbour (Malaysia), 1.2 cm.y^{-1} reported by Tang et al. (2008) for the Victoria Harbour (Hong Kong) and 0.2-2 cm.y^{-1} reported by Smith et al. (2009) for the Sydney Harbour (Nova Scotia, Canada).

9.1.2.2. Ecotoxicological implications of metal contamination in settling silt-clay particles in the Tema Harbour

The fine fractions of sediments, particularly the silt-clay particles, can be easily ingested by benthic organisms (Bat and Raffaelli, 1998; Bat, 2005). Therefore, chemical contamination of silt-clay particles is a threat to benthic organisms and other organisms in the aquatic food chain. Based on the ERL and ERM sediment quality guidelines (Long et al., 1995; Long et al., 1998), Chapter 8 assessed the ecotoxicological implications of metal contamination in settling silt-clay particles collected in the sediment traps (Chapter 7). The results showed that As concentrations in the trapped silt-clay particles (146-1470 mg.kg^{-1} dw; see Table 8.1) far exceeded its ERM value (70 mg.kg^{-1} dw; Table 8.7) by about 2-3 orders of magnitude. In most cases, the concentrations of Cr (86%), Cu (86%), Zn (93%), Ni (71%) and Hg (77%) in the trapped silt-clay particles were higher than their respective ERL values, but lower than their corresponding ERM values (Table 8.7). Thus, the settling silt-clay particles were potentially hazardous and pose medium to high ecotoxicological risks to benthic invertebrates due to the concentrations of As, Cr, Cu, Zn, Ni and Hg.

The As concentrations in the settling silt-clay particles were higher than previously measured As concentrations in bulk surface sediments (Table 5.1) or As concentrations in silt-clay normalised sediments from the Tema Harbour (Table 6.6). Table 8.6 showed that the As concentrations in the settling silt-clay particles were high compared with As concentrations reported for sediments from gold mining areas in Ghana (Affum et al., 2016) and coastal areas of Egypt (Mostafa et al., 2004; Abdel Ghani et al., 2013), Finland (Vallius et al., 2007), China (Luo et al., 2010), Spain (Díaz-de Alba et al., 2011), Tanzania (Rumisha et al., 2012), the Mediterranean (Mamindy-Pajany et al., 2013), and Saudi Arabia (Al-Taani et al., 2014; Youssef et al., 2015; El-Sorogy et al., 2016). An increasing trend in As concentrations in the settling silt-clay particles was observed across the sampling period (Fig. 8.1i), which suggested the existence of continuous sources of As at the Tema Harbour. Potential As sources include industrial activities such as metal smelting, the use of As-containing agrochemicals in the harbour area and spillage of As-containing chemicals in the harbour.

9.2. General conclusions

This study is the first comprehensive assessment of sediments in the Tema Harbour in terms of chemical pollution and accumulation rates. The thesis contributes to improving our ability to combine sediment chemistry with bioassays in one comprehensive assessment of the contamination state of a coastal harbour. The study results showed that the harbour has been severely affected by anthropogenic activities, resulting in pollution of the sediments, especially those from the Fishing Harbour and the Canoe Basin. This raises concerns about sediment and general environmental management at the Tema Harbour, considering the lack of engineered fields where dredged materials from the harbour can be stored or treated. The expansion of the Tema Harbour and maintenance operations requires dredging and disposal of the dredged material at sea may be the only option. In the context of this study, there is a need to undertake remedial measures that will minimize the impact of dredging and the disposal of dredged sediments from the harbour. Equally important is the need for enforcement of environmental regulations to control point and non-point sources of pollution at the Tema Harbour.

This thesis further demonstrates how sediment trap data can complement sediment core data with the use of multi-tracers and radiometric dating models in the marine environment to obtain reliable information on sediment dynamics in a disturbed coastal harbour area where conventional ^{210}Pb-based dating methods fail. In particular, the ^{210}Pb-based TERESA model is a good tool for quantifying sedimentation rates in the Tema Harbour with overall time-averaged SAR values in the range of 1.4-3.0 g.cm^{-2}.y^{-1} and sediment accretion rates of 1.7-3 cm.y^{-1}. These accretion rates pose moderate problems for sustainable management of the harbour. The main findings of the thesis for environmental management at the Tema Harbour are presented in Fig.9.1.

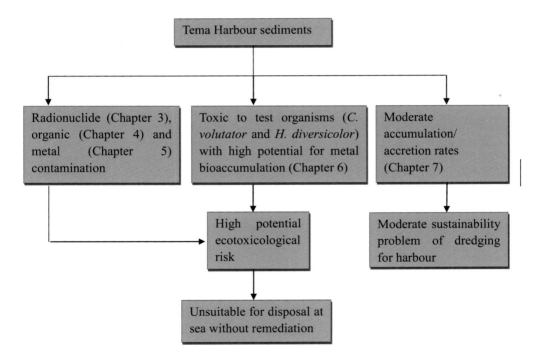

Fig. 9.1 Schematic presentation of the main findings of the thesis for environmental management at the Tema Harbour

References

Abdel Ghani, S., El Zokm, G., Shobier, A., Othman, T., & Shreadah, M. (2013). Metal pollution in surface sediments of Abu-Qir Bay and Eastern Harbour of Alexandria, Egypt. *The Egyptian Journal of Aquatic Research, 39*(1), 1-12.

Abrahim, G.M.S. & Parker, R.J. (2008). Assessment of heavy metal enrichment factors and the degree of contamination in marine sediments from Tamaki Estuary, Auckland, New Zealand. *Environ Monit Assess. 136*, 227-238

Abrajano Jr, T. A., Yan, B., Song, J., Bopp, R., & O'Malley, V. (2007). High-Molecular-Weight Petrogenic and Pyrogenic Hydrocarbons in Aquatic Environments. In D. H. Editors-in-Chief: Heinrich & K. T. Karl (Eds.), *Treatise on Geochemistry* (pp. 1-50). Oxford: Pergamon.

Adams, M.S. & Stauber, J.L. (2008). Marine whole sediment toxicity tests for use in temperate and tropical Australian environments: current status. *Aust. J. Ecotoxicol. 14*, 155-167.

Adams, W.J., Blust, R., Borgmann, U., Brix, K.V., DeForest, D.K., Green, A.S., Meyer, J.S., McGeer, J.C., Paquin, P.R., & Rainbow, P.S. (2011). Utility of tissue residues for predicting effects of metals on aquatic organisms. *Integr. Environ. Assess. Manage. 7*, 75-98.

Addo, M.A., Affum, H., Botwe, B.O., Gbadago, J.K., Acquah, S.A., Senu, J.K., Adom, T. & Coleman, A. (2012). Assessment of Water Quality and Heavy Metal Levels in Water and Bottom Sediment Samples from Mokwé Lagoon, Accra, Ghana. *Res. J. Environ. Earth Sci., 4*(2), 119-130.

Affum, A.O., Osae Dede, S., Nyarko, B.J.B, Acquaah, S., Kwaansa-Ansah, E.E, Darko, G., Dickson, A., Affum, E., & Fianko, J.R. (2016). Influence of small-scale gold mining and toxic element concentrations in Bonsa river, Ghana: A potential risk to water quality and public health. *Environ. Earth Sci., 75*, 178-194.

Akrasi, S. (2011). Sediment discharges from Ghanaian rivers into the sea. *West Afr. J. Appl. Ecol. 18*, 1-13.

Al-Taani, A.A., Batayneh, A., Nazzal, Y., Ghrefat, H., Elawadi, E., & Zaman, H. (2014). Status of trace metals in surface seawater of the Gulf of Aqaba, Saudi Arabia. *Marine Pollution Bulletin, 86*(1), 582-590.

Aydin-Onen, S., Kucuksezgin, F., Kocak, F., & Açik, S. (2015). Assessment of heavy metal contamination in *Hediste diversicolor* (OF Müller, 1776), Mugil cephalus (Linnaeus,

1758), and surface sediments of Bafa Lake (Eastern Aegean). *Environ. Sci. Pollut. Res. 22*, 8702-8718.

Bat, L. (2005). A review of sediment toxicity bioassays using the amphipods and polychaetes. *Turkish Journal of Fisheries and Aquatic Sciences, 5*(2), 119-139.

Bat, L. & Raffaelli, D. (1998). Sediment toxicity testing: a bioassay approach using the amphipod *Corophium volutator* and the polychaete *Arenicola marina. J. Exp. Mar. Bio. Ecol. 226*, 217-239.

Bellucci, L. G., Giuliani, S., Romano, S., Albertazzi, S., Mugnai, C., & Frignani, M. (2012). An Integrated Approach to the Assessment of Pollutant Delivery Chronologies to Impacted Areas: Hg in the Augusta Bay (Italy). *Environ. Sci. & Technol. 46*(4), 2040-2046.

Birch, G.F., & Hutson, P., 2009. Use of sediment risk and ecological/conservation value for strategic management of estuarine environments: Sydney estuary, Australia. *Environ. Manage. 44*, 836-850.

Birch, G.F. & Olmos, M.A. (2008). Sediment-bound heavy metals as indicators of human influence and biological risk in coastal water bodies. ICES *J. Mar. Sci. 65*: 1407-1413.

Bolam, S.G., Rees, H.L., Somerfield, P., Smith, R., Clarke, K.R., Warwick, R.M., Atkins, M. & Garnacho, E. (2006). Ecological consequences of dredged material disposal in the marine environment: a holistic assessment of activities around the England and Wales coastline. *Mar. Pollut. Bull., 52*(4), 415-26.

Buesseler, K.O., Antia, A.N., Chen, M., Fowler, S.W., Gardner, W.D., Gustafsson, O., Harada, K., Michaels, A.F., Rutgers van der Loeff, M. & Sarin, M. (2007). An assessment of the use of sediment traps for estimating upper ocean particle fluxes. J Mar Res *65*(3), 345-416.

Burton, G.A. (2002). Sediment quality criteria in use around the world. *Limnology, 3*, 65-76.

Caille, N., Tiffreau, C., Leyval, C. & Morel, J.L., 2003. Solubility of metals in an anoxic sediment during prolonged aeration. *Sci. Total Environ. 301*(1), 239-250.

Carvalho, S., Cunha, M.R., Pereira, F., Pousão-Ferreira, P., Santos, M., & Gaspar, M. (2012). The effect of depth and sediment type on the spatial distribution of shallow soft-bottom amphipods along the southern Portuguese coast. *Helgol. Mar. Res. 66*, 489-501.

Casado-Martínez MC., Buceta JL, Belzunce MJ, DelValls TA (2006) Using sediment quality guidelines for dredged material management in commercial ports from Spain. *Environ. Int. 32*, 388-396.

Casado-Martinez, M.C., Forja, J.M., & DelValls, T.A., 2007. Direct comparison of amphipod sensitivities to dredged sediments from Spanish ports. *Chemosphere, 68*, 677-685.

Choueri, R.B., Cesar, A., Abessa, D.M.S., Torres, R.J., Morais, R.D., Riba, I, Pereira, C.D.S., Nascimento, M.R.L., Mozeto, A.A. & DelValls, T.A. (2009). Development of site-specific sediment quality guidelines for North and South Atlantic littoral zones: Comparison against national and international sediment quality benchmarks. *J. Hazard. Mater. 170*, 320-331.

Ciarelli, S., Vonck, W., Van Straalen, N., & Stronkhorst, J., 1998. Ecotoxicity assessment of contaminated dredged material with the marine amphipod *Corophium volutator. Arch. Environ. Contam. Toxicol. 34*, 350-356.

Dauvin, J.C., Andrade, H., de-la-Ossa-Carretero, J.A., Del-Pilar-Ruso, Y., Riera, R. (2016). Polychaete/amphipod ratios: An approach to validating simplebenthic indicators. *Ecological Indicators, 63*, 89-99.

DelValls, T., Andres, A., Belzunce, M., Buceta, J., Casado-Martinez, M., Castro, R., Riba, I., Viguri, J.R. & Blasco, J. (2004). Chemical and ecotoxicological guidelines for managing disposal of dredged material. *Trends Anal. Chem. 23*, 819-828.

Díaz-Asencio, M., Alonso-Hernández, C. M., Bolanos-Álvarez, Y., Gómez-Batista, M., Pinto, V., Morabito, R., . . . Sanchez-Cabeza, J. A. (2009). One century sedimentary record of Hg and Pb pollution in the Sagua estuary (Cuba) derived from ^{210}Pb and ^{137}Cs chronology. *Marine Pollution Bulletin, 59*(4–7), 108-115.

Díaz-de Alba, M., Galindo-Riaño, M. D., Casanueva-Marenco, M. J., García-Vargas, M. & Kosore, C. M. (2011). Assessment of the metal pollution, potential toxicity and speciation of sediment from Algeciras Bay (South of Spain) using chemometric tools. *J. Hazard. Mater. 190*(1-3), 177-187.

Dung, T.T.T., Cappuyns, V., Swennen, R. & Phung, N.K. (2013). From geochemical background determination to pollution assessment of heavy metals in sediments and soils. *Rev Environ Sci Bio/Technol*, 12, 335-353.

El-Sorogy, A.S., Youssef, M., Al-Kahtany, K., & Al-Otaiby, N. (2016). Assessment of arsenic in coastal sediments, seawaters and molluscs in the Tarut Island, Arabian Gulf, Saudi Arabia. *J. Afr. Earth Sci., 113*, 65-72.

EPA, USACE (1998). Evaluation of dredged material proposed for discharge in waters of the US-testing manual: Inland Testing Manual. EPA-823-B-98-004. US Environmental Protection Agency and US Army Corps of Engineers, Washington, DC.

Erten, H., Von Gunten, H., Rössler, E. & Sturm, M. (1985). Dating of sediments from Lake Zurich (Switzerland) with210Pb and [137]Cs. *Swiss J Hydrol*, *47*(1), 5-11.

Erten H. (1997). Radiochronology of lake sediments. *Pure Appl Chem*, *69*(1),71-76.

Escher, B.I., Bramaz, N., Mueller, J.F., Quayle, P., Rutishausera, S. & Vermeirssena, E.L.M (2008). Toxic equivalent concentrations (TEQs) for baseline toxicity and specific modes of action as a tool to improve interpretation of ecotoxicity testing of environmental samples. *J. Environ. Monit*. *10*, 612-621.

Forrester, G.E., Fredericks, B.I., Gerdeman, D., Evans, B., Steele, M.A., Zayed, K., Schweitzer, L.E., Suffet, I.H., Vance, R.R., & Ambrose, R.F., 2003. Growth of estuarine fish is associated with the combined concentration of sediment contaminants and shows no adaptation or acclimation to past conditions. *Mar. Environ. Res*. *56*, 423-442.

Green, M.O. & Coco, G. (2014). Review of wave-driven sediment resuspension and transport in estuaries. *Rev. Geophys*. *52*(1), 77-117.

Guerra, R. (2012). Polycyclic Aromatic Hydrocarbons, Polychlorinated Biphenyls and Trace Metals in Sediments from a Coastal Lagoon (Northern Adriatic, Italy). *Water, Air, & Soil Pollut. 223*(1), 85-98.

Hannewijk, A., Kater, B.J. & Schipper, C.A. (2004). Sediment toxiciteitstest met *Nereis diversicolor* (In Dutch). 22 pp.

Hong, H., Xu, L.-J., Zhang, L., Chen, J., Wong, Y. & Wan, T. (1995). Special guest paper: environmental fate and chemistry of organic pollutants in the sediment of Xiamen and Victoria Harbours. *Mar. Pollut. Bull. 31*(4), 229-236.

Jain, C. (2004). Metal fractionation study on bed sediments of River Yamuna, India. *Water Res.*, 38, 569-578.

Jiang, J., Wang, J., Liu, S., Lin, C, He, M. & Liu, X. (2013). Background, baseline, normalization, and contamination of heavy metals in the Liao River Watershed sediments of China. *J. Asian Earth Sci. 73*, 87-94.

Kalwa, M., Quináia, S., Pletsch, A., Techy, L., & Felsner, M. (2013). Fractionation and Potential Toxic Risk of Metals From Superficial Sediment in Itaipu Lake—Boundary Between Brazil and Paraguay. *Arch. Environ. Contam. Toxicol. 64*(1), 12-22.

Kelderman P, Ang'weya R, De Rozari P, Vijverberg T. (2012). Sediment characteristics and wind-induced sediment dynamics in shallow Lake Markermeer, the Netherlands. Aquat Sci *74*(2), 301-313.

Klubi, E., Abril, J.M., Nyarko, E., Laissaoui, A. & Benmansour, M. (2017). Radioecological assessment and radiometric dating of sediment cores from dynamic sedimentary systems of Pra and Volta estuaries (Ghana) along the Equatorial Atlantic. *J. Environ. Radioact.*, *178-179*, 1-11.

Lepland, A., Andersen, T.J., Lepland, A., Arp, H.P.H., Alve, E., Breedveld, G.D., & Rindby, A., 2010. Sedimentation and chronology of heavy metal pollution in Oslo harbour, Norway. *Mar. Pollut. Bull.* *60*, 1512-1522.

Leys, V., & Mulligan, R.P. (2011). Modelling coastal sediment transport for harbour planning: selected case studies. INTECH.

Long, E.R., Ingersoll, C.G. & MacDonald, D.D. (2006). Calculation and uses of mean sediment quality guideline quotients: a critical review. *Environ. Sci. Technol. 40*(6), 1726-1736.

Long, E.R., Field, L.J., & MacDonald, D.D. (1998). Predicting toxicity in marine sediments with numerical sediment quality guidelines. *Environ. Toxicol. Chem.*, *17*(4), 714-727.

Long, E.R., MacDonald, D.D., Smith, S. L., & Calder, F.D. (1995). Incidence of adverse biological effects within ranges of chemical concentrations in marine and estuarine sediments. *Environ. Manage.*, *19*(1), 81-97.

Lu, X. (2007). A note on removal of the compaction effect for the ^{210}Pb method. *Appl. Radiat. Isot, 65*(1), 142-146.

Luo, W., Lu, Y., Wang, T., Hu, W., Jiao, W., Naile, J.E., Khim, J.S. & Giesy, J.P. (2010). Ecological risk assessment of arsenic and metals in sediments of coastal areas of northern Bohai and Yellow Seas, China. *Ambio, 39*(5-6), 367-375.

Luoma, S.N. & Rainbow, P.S. (2008). Metal contamination in aquatic environments: science and lateral management. Cambridge University Press, Cambridge.

Mahu, E, Nyarko, E., Hulme, S. & Coale, K.H. (2015). Distribution and enrichment of trace metals in marine sediments from the Eastern Equatorial Atlantic, off the Coast of Ghana in the Gulf of Guinea. *Mar. Pollut. Bull. 98*, 301-307.

Mamindy-Pajany, Y., Bataillard, P., Séby, F., Crouzet, C., Moulin, A., Guezennec, A.-G., Hrel, C., Marmier, N., & Battaglia-Brunet, F. (2013). Arsenic in Marina Sediments from the Mediterranean Coast: Speciation in the Solid Phase and Occurrence of Thioarsenates. *Soil Sed Contam: An Int. J., 22*(8), 984-1002.

Mayor, D.J., Solan, M., Martinez, I., Murray, L., McMillan, H., Paton, G.I., & Killham, K., (2008). Acute toxicity of some treatments commonly used by the salmonid aquaculture industry to *Corophium volutator* and *Hediste diversicolor*: Whole sediment bioassay tests. *Aquaculture, 285*, 102-108.

Mestres M, Sierra J, Mösso C. & Sánchez-Arcilla, A (2010). Sources of contamination and modelled pollutant trajectories in a Mediterranean harbour (Tarragona, Spain). *Mar. Pollut. Bull. 60*, 898-907.

Moreira, S.M., Lima, I., Ribeiro, R., & Guilhermino, L. (2006). Effects of estuarine sediment contamination on feeding and on key physiological functions of the polychaete *Hediste diversicolor*: laboratory and in situ assays. *Aquat Toxicol. 78*, 186-201.

Mostafa, A.R., Barakat, A.O., Qian, Y., Wade, T.L., & Yuan, D. (2004). An overview of metal pollution in the Western Harbour of Alexandria, Egypt. *Soil Sed Contam: An Int. J., 13*(3), 299-311.

Muller, G., 1969. Index of geo-accumulation in sediments of the Rhine River. *J. Geol., 2*(3): 108-118.

Mulligan, C.N., Yong, R.N. & Gibbs, B.F. (2001). An evaluation of technologies for the heavy metal remediation of dredged sediments. *J. Hazard. Mater. 85*, 145-163.

Nyarko, E., Botwe, B.O. & Klubi, E. (2011). Polycyclic aromatic hydrocarbons (PAHs) levels in two commercially important fish species from coastal waters of Ghana and their carcinogenic health risks. *West Afr. J. Appl. Ecol., 19*, 53-66.

Petrosillo, I., Valente, D., Zaccarelli, N. & Zurlini, G. (2009). Managing tourist harbors: Are managers aware of the real environmental risks? *Mar. Pollut. Bull. 58*, 1454-1461.

Pfitzner, J., Brunskill, G. & Zagorskis, I. (2004). ^{137}Cs and excess ^{210}Pb deposition patterns in estuarine and marine sediment in the central region of the Great Barrier Reef Lagoon, north-eastern Australia. *J. Environ. Radioact. 76*(1), 81-102.

Philippe, S., Leterme, C., Lesourd, S., Courcot, L., Haack, U. & Caillaud, J. (2008). Bioavailability of sediment-borne lead for ragworms (*Hediste diversicolor*) investigated by lead isotopes. *Appl. Geochem. 23*, 2932-2944.

Pini J, Richir J, Watson G (2015) Metal bioavailability and bioaccumulation in the polychaete Nereis (Alitta) virens (Sars): The effects of site-specific sediment characteristics. *Mar. Pollut. Bull., 95*, 565-575.

Prato, E., Parlapiano, I. & Biandolino, F. (2011). Evaluation of a bioassays battery for ecotoxicological screening of marine sediments from Ionian Sea (Mediterranea Sea, Southern Italy). *Environ Monit Assess*, DOI 10.1007/s10661-011-2335-9.

Rainbow, P.S. & Luoma, S.N. (2011). Metal toxicity, uptake and bioaccumulation in aquatic invertebrates – Modelling zinc in crustaceans. *Aquat Toxicol. 105*, 455-465

Rainbow, P.S., Smith, B.D. & Luoma, S.N. (2009). Differences in trace metal bioaccumulation kinetics among populations of the polychaete *Nereis diversicolor* from metal-contaminated estuaries. *Mar Ecol Prog Ser.*, *376*, 173-184.

Ré, A., Freitas, R., Sampaio, L., Rodrigues, A.M., & Quintino, V. (2009). Estuarine sediment acute toxicity testing with the European amphipod *Corophium multisetosum* Stock, 1952. *Chemosphere*, *76*, 1323-1333.

Rocher, V., Azimi, S., Moilleron, R., & Chebbo, G. (2004). Hydrocarbons and heavy metals in the different sewer deposits in the 'Le Marais' catchment (Paris, France): stocks, distributions and origins. *Sci. Total Environ., 323*(1-3), 107-122.

Roddie, B., & Thain, J. (2002). Biological effects of contaminants: *Corophium sp.* sediment bioassay and toxicity test: International Council for the Exploration of the Sea. ICES Techniques in Marine Environmental Sciences No. 28, ISSN 0903-2606, Copenhagen, .Denmark.

Romero, A.F., Asmus, M.L., Milanelli, J. C. C., Buruaema, L., Abessa, D.M.S. (2014). Self-diagnosis method as an assessment tool for environmental management of Brazilian ports. *J. Integr. Coast. Zone Manage., 14*(4):637-644.

Rose, A., Carruthers, A-M, Stauber, J., Lima, R & Blockwell, S. (2006). Development of an acute toxicity test with the marine copepod Acartia Sinjiensis. *Aust. J. Ecotoxicol, 12*, 73-81.

Ruiz-Fernández, A.C., Frignani, M., Hillaire-Marcel, C., Ghaleb, B., Arvizu, M.D., Raygoza-Viera, J.R. & Páez-Osuna, F. (2009). Trace metals (Cd, Cu, Hg, and Pb) accumulation recorded in the intertidal mudflat sediments of three coastal lagoons in the Gulf of California, Mexico. *Estuar. Coasts.* DOI 10.1007/s12237-009-9150-3.

Rumisha, C., Elskens, M., Leermakers, M., & Kochzius, M. (2012). Trace metal pollution and its influence on the community structure of soft bottom molluscs in intertidal areas of the Dar es Salaam coast, Tanzania. *Marine Pollution Bulletin, 64*(3), 521-531.

Santschi, P.H., Presley, B.J., Wade, T.L., Garcia-Romero, B., & Baskaran, M. (2001). Historical contamination of PAHs, PCBs, DDTs, and heavy metals in Mississippi River Delta, Galveston Bay and Tampa Bay sediment cores. *Mar. Environ. Res. 52*(1), 51-79.

Scaps, P. (2002). A review of the biology, ecology and potential use of the common ragworm *Hediste diversicolor* (OF Müller)(Annelida: Polychaeta). *Hydrobiologia, 470*, 203-218.

Scarlett, A., Rowland, S., Canty, M., Smith, E., & Galloway, T. (2007). Method for assessing the chronic toxicity of marine and estuarine sediment-associated contaminants using the amphipod *Corophium volutator*. *Mar. Environ. Res. 63*, 457-470.

Schipper, C., Rietjens, I., Burgess, R., Murk, A. (2010). Application of bioassays in toxicological hazard, risk and impact assessments of dredged sediments. *Mar. Pollut. Bull. 60*(11), 2026-2042.

Schipper, C.A., Burgess, R.M., van den Dikkenberg, L.C. (2006). Sediment toxiciteittest met de slijkgarnaal *Corophium volutator* (In Dutch). 18 pp.

Schmidt S, Jouanneau J-M, Weber O, Lecroart P, Radakovitch O, Gilbert F, Jézéquel D. (2007a). Sedimentary processes in the Thau Lagoon (France): from seasonal to century time scales. *Estuar Coast Shelf Sci*; *72*(3), 534-542.

Schmidt S, Gonzalez J-L, Lecroart P, Tronczyński J, Billy I, Jouanneau J-M. (2007b). Bioturbation at the water-sediment interface of the Thau Lagoon: impact of shellfish farming. Aquat Living Resour; *20*(2), 163-169.

Sinex, S.A. & Helz, G.R. (1981). Regional Geochemistry of Trace Elements in Chesapeake Bay Sediments. *Environ. Geol. 3*, 315-323.

Smith, J., Lee, K., Gobeil, C. & Macdonald, R. (2009). Natural rates of sediment containment of PAH, PCB and metal inventories in Sydney Harbour, Nova Scotia. *Sci. Total Environ. 407*(17), 4858-4869.

Smoak, J.M. & Patchineelam, S.R. (1999). Sediment mixing and accumulation in a mangrove ecosystem: evidence from ^{210}Pb, [234]Th and [7]Be. *Mangroves and Salt Marshes*, *3*, 17-27.

Smoak, J.M., Moore, W.S., Thunell, R.C. & Shaw, T.J. (1999). Comparison of [234]Th, [228]Th, and ^{210}Pb fluxes with fluxes of major sediment components in the Guaymas Basin, Gulf of California. *Mar. Chem.*, *65*, 177-194.

Stronkhorst, J. & Van Hattum, B. (2003). Contamination of concern in Dutch marine harbour sediments. *Arch. Environ. Contam. Toxicol.*, *45*, 306-316.

Stronkhorst, J., Schipper, C.A., Brils, J., Dubbeldam, M., Postma, J. & van de Hoeven, N. (2003). Using marine bioassays to classify the toxicity of Dutch harbour sediments. *Environ Toxicol Chem.*, *22* (7), 1535-1547.

Su, C.-C., & Huh, C.-A. (2002). 210Pb, 137Cs and 239,240Pu in East China Sea sediments: sources, pathways and budgets of sediments and radionuclides. *Mar. Geol.*, *183*(1–4), 163-178.

Syvitski, J.P., Vörösmarty, C.J., Kettner, A.J. & Green, P. (2005). Impact of humans on the flux of terrestrial sediment to the global coastal ocean. *Science*, *308*(5720), 376-380.

Tang, C. W.-y., Ip, C. C.-m., Zhang, G., Shin, P. K. S., Qian, P.-y., & Li, X.-d. (2008). The spatial and temporal distribution of heavy metals in sediments of Victoria Harbour, Hong Kong. *Mar. Pollut. Bull.*, *57*(6-12), 816-825.

Taylor, S. (1964). Abundance of chemical elements in the continental crust: a new table. *Geochim Cosmochim Acta*, *28*, 1273-1285.

Uwadiae, R.E. (2010). An inventory of the benthic macrofauna of Epe Lagoon, southwest Nigeria. *J. Sci. Res. Develop. 12*, 161-171.

Vallius, H., Ryabchuk, D., Kotilainen, A. (2007). Distribution of heavy metals and arsenic in soft surface sediments of the coastal area off Kotka, northeastern Gulf of Finland, Baltic Sea. In: Vallius, H. (Ed.), Holocene Sedimentary Environment and Sediment Geochemistry of the Eastern Gulf of Finland, Baltic Sea, Geological Survey, Finland, 45:33-48.

van Dam, R.A., Harford, A.J., Houston, M.A., Hogan, A.C. & Negri, A.P. (2008). Tropical marine toxicity testing in australia: a review and recommendations. *Aust. J. Ecotoxicol.*, *14*, 55-88.

van den Heuvel-Greve, M., Postma, J., Jol, J., Kooman, H., Dubbeldam, M., Schipper, C., & Kater, B. (2007). A chronic bioassay with the estuarine amphipod *Corophium volutator*: Test method description and confounding factors. *Chemosphere*, *66*, 1301-1309.

van Hullebusch, E.D., Lens, P.N.L. & Tabak, H.H. (2005). Developments in bioremediation of soils and sediments polluted with metals and radionuclides. 3. Influence of chemical speciation and bioavailability on contaminants immobilization/mobilization bio-processes. *Rev. Environ. Sci. Biotechnol.*, *4*(3), 185–212.

Witt, J., Schroeder, A., Knust, R. & Arntz, W.E. (2004). The impact of harbour sludge disposal on benthic macrofauna communities in the Weser estuary. *Helgol. Mar. Res.* 58, 117-128.

Yeager, K., Santschi, P., Phillips, J. & Herbert, B. (2005). Suspended sediment sources and tributary effects in the lower reaches of a coastal plain stream as indicated by radionuclides, Loco Bayou, Texas. *Environ. Geol. 47*(3), 382-395.

Yilgor, S., Kucuksezgin, F., & Ozel, E. (2012). Assessment of metal concentrations in sediments from Lake Bafa (Western Anatolia): an index analysis approach. *Bulletin of Environ. Contam. Toxicol.*, *89*(3), 512-518.

Youssef, M., El-Sorogy, A., Al Kahtany, K., & Al Otiaby, N. (2015). Environmental assessment of coastal surface sediments at Tarut Island, Arabian Gulf (Saudi Arabia). *Marine Pollution Bulletin, 96*(1), 424-433.

Yunker, M.B., Macdonald, R.W., Vingarzan, R., Mitchell, R.H., Goyette, D. & Sylvestre, S. (2002). PAHs in the Fraser River basin: a critical appraisal of PAH ratios as indicators of PAH source and composition. *Org. Geochem. 33*(4), 489-515.

Yusoff, A.H., Zulkifli, S.Z., Ismail, A. & Mohamed, C.A.R. (2015). Vertical trend of trace metals deposition in sediment core off Tanjung Pelepas Harbour, Malaysia. *Procedia Environ Sci, 30*, 211-216.

Samenvatting

De haven van Tema in Ghana is al bijna zes decennia in bedrijf en is onderhevig aan een grote toevoer van sedimenten en verontreinigingde sedimenten als gevolg van de intense antropogene activiteiten in het havengebied. Dit proefschrift beoordeelde de sediment verontreiniging in de Tema haven door gebruik te maken van de standaard 10-daagse *Corophium volutator* en 28-daagse *Hediste diversicolor* sediment toxiciteit bioassays evenals het bepalen van de concentraties van chemische contaminanten (DDT's, HCH's, PAK's en metaal - Cd, Pb, Cr, Ni , Cu, Zn en As). De resultaten van de bioassays toonden significante *C. volutator* en *H. diversicolor* mortaliteit, wat aangeeft dat de Tema Harbour-sedimenten vervuild en toxisch zijn. Biota-sediment accumulatiefactoren onthulden verder een hoog potentieel voor bioaccumulatie van de sediment-geassocieerde metalen, wat nadelige gevolgen kan hebben voor de voedselketen. De sedimenten van de Tema haven zijn dus ongeschikt voor baggeren en dumpen op zee zonder sanering.

Het proefschrift onderzocht verder de accumulatie percentages van sedimenten (SAR's) in de Tema-haven door de gecombineerde analyse van sediment profielen en sediment profiel gegevens. De sediment profielen vertoonden variabele bulk dichtheidsprofielen, wat wijst op zeer dynamische en niet-stabiele sedimentatie omstandigheden. De brutoschattingen van zeer recente sediment accumulatiesnelheden met behulp van het constante flux sedimentatie (CF-CS) model lagen in het bereik van 2,5-9,0 $g.cm^{-2}.y^{-1}$. Deze waarden waren veel lager dan de geschatte gemiddelde bezinkingsfluxen op basis van de gegevens uit sedimentenvallen (15,2-53,8 $g.cm^{-2}.y^{-1}$), wat aangeeft dat resuspensie van de sedimenten een belangrijke rol speelt in sedimentatie in de Tema haven. Conventionele ^{210}Pb modellen voor sediment datering konden niet worden gebruikt om de SAR's in de Tema-haven te schatten. Het TERESA-model op basis van ^{210}Pb bleek echter een goed hulpmiddel te zijn om de accumulatie van sedimenten in de Tema haven te kwantificeren met tijdsgemiddelde waarden variërend tussen 1,4-3,0 $g.cm^{-2}.y^{-1}$ en met sediment accumulatiesnelheden van 1,7-3 $cm.y^{-1}$.

Concluderend heeft deze studie aangetoond dat de haven van Tema ernstig is aangetast door antropogene activiteiten, resulterend in vervuiling van de sedimenten, vooral die van de Visserijhaven en het Kanobasin. Bovendien kan de sediment accumulatie in de haven gematigde problemen opleveren voor het duurzaam gebruik van de haven. Er is daarom behoefte aan verbetering van het sediment en milieubeheer in de haven van Tema en het reguleren van de afvoer van het gebaggerde materiaal uit deze tropische kusthaven.

Netherlands Research School for the
Socio-Economic and Natural Sciences of the Environment

D I P L O M A

For specialised PhD training

The Netherlands Research School for the
Socio-Economic and Natural Sciences of the Environment
(SENSE) declares that

Benjamin Botwe

born on 25 June 1974 in Accra, Ghana

has successfully fulfilled all requirements of the
Educational Programme of SENSE.

Delft, 29 June 2018

the Chairman of the SENSE board

Prof. dr. Huub Rijnaarts

the SENSE Director of Education

Dr. Ad van Dommelen

The SENSE Research School has been accredited by the Royal Netherlands Academy of Arts and Sciences (KNAW)

The SENSE Research School declares that **Benjamin Botwe** has successfully fulfilled all
requirements of the Educational PhD Programme of SENSE with a
work load of 33.2 EC, including the following activities:

SENSE PhD Courses

o   Environmental research in context (2014)
o   Research in context activity: 'Co-organizing UNESCO-IHE PhD Symposium on: From
    water scarcity to water security', 03-04 October 2016, Delft, The Netherlands

Other PhD and Advanced MSc Courses

o   Sampling in space and time for survey and monitoring of natural resources, PE&RC
    graduate school, Wageningen University (2013)
o   Speciation and Bioavailability of Metals, Organics and Nanoparticles, Wageningen
    University (2013)
o   Summer School on Contaminated Sediments - Characterization and Remediation, IHE
    Delft (2016)

Research Supporting Activities

o   Extended literature review
o   Several published papers

Oral Presentations

o   *Short-term sediment-associated trace metal dynamics in coastal marine Tema Harbour
    (Ghana).* YOUMARES 7 Conference, 11-13 September 2016, Hamburg, Germany
o   *Settling fluxes and ecotoxicological risk assessment of fine sedimentary metals in Tema
    Harbour (Ghana).* Conference on fisheries and coastal environment, 25-27 September
    2017, Accra, Ghana

SENSE Coordinator PhD Education

Dr. Peter Vermeulen

For Product Safety Concerns and Information please contact our EU
representative GPSR@taylorandfrancis.com Taylor & Francis Verlag GmbH,
Kaufingerstraße 24, 80331 München, Germany

Printed and bound by CPI Group (UK) Ltd, Croydon, CR0 4YY
01/05/2025
01858616-0002